PLANE SENSE

A Beginner's Guide to Owning and Operating Private Aircraft

FAA-H-8083-19A

U.S. Department of Transportation

FEDERAL AVIATION ADMINISTRATION

Flight Standards Service

Skyhorse Publishing

Skyhorse Publishing books may be purchased in bulk at special discounts for sales promotion, corporate gifts, fund-raising, or educational purposes. Special editions can also be created to specifications. For details, contact the Special Sales Department, Skyhorse Publishing, 307 West 36th Street, 11th Floor, New York, NY 10018 or info@skyhorsepub-lishing.com.

www.skyhorsepublishing.com

10 9 8 7 6 5 4 3 2

Library of Congress Cataloging-in-Publication Data

Plane sense : a beginner's guide to owning and operating private aircraft.
p. cm.
ISBN 978-1-61608-133-1 (pbk. : alk. paper)
1. Private planes.
TL685.1.P525 2010
629.132'5--dc22

2010033652

Printed in China

Preface

Plane Sense introduces aircraft owners and operators, or prospective aircraft owners and operators, to basic information about the requirements involved in acquiring, owning, operating, and maintaining a private aircraft.

This handbook can be a valuable reference tool for anyone who would like to review the "nuts and bolts" of aircraft ownership. Aircraft owners and operators, or anyone considering aircraft ownership, should be familiar with Title 14 of the Code of Federal Regulations (14 CFR), which details regulations for aircraft owners, operators, pilots, aircraft mechanics, and maintenance providers. Since the requirements can be updated and the regulations can change, the Federal Aviation Administration (FAA) recommends that you contact your nearest FAA Flight Standards District Office (FSDO), where the personnel can assist you with the various requirements for aircraft ownership, operation, and maintenance.

The FAA has also added information for aviation enthusiasts who own (or are interested in owning) light-sport aircraft, a new and evolving sector of the general aviation marketplace.

This handbook highlights regulations and regulatory guidance material, as well as providing advice regarding where to locate answers to your questions. While *Plane Sense* cannot cover every issue faced by aircraft owners and operators, this handbook is intended to be a useful guide and will help you locate the resources to assist you.

This publication supersedes FAA-8083-19, which was reprinted with editorial updates in 2003.

This handbook is available free of charge for download, in PDF format, from the FAA Regulatory Support Division (AFS-600) on the FAA website at www.faa.gov.

Plane Sense may be also be purchased from:

Superintendent of Documents
United States Government Printing Office
Washington, DC 20402-9325
http://bookstore.gpo.gov

This handbook is published by and comments should be sent to:

Federal Aviation Administration
Airman Testing Standards Branch (AFS-630)
P.O. Box 25082
Oklahoma City, OK 73125
afs630comments@faa.gov

Acknowledgments

Plane Sense was produced by the Federal Aviation Administration (FAA) with the assistance of The Wicks Group, PLLC. The FAA wishes to acknowledge the providers of the following images used in this handbook:

Skycatcher used on the cover and in chapter 6, courtesy of Cessna Aircraft Corporation

Cirrus SR20-1 used on the cover and in chapter 1, courtesy of Cirrus Design

Eclipse 500 used on the cover, courtesy of Eclipse Aviation Corporation

The FAA would also like to extend its appreciation to several aviation industry organizations that provided assistance and input in the preparation of this handbook, including:

General Aviation Manufacturers Association (GAMA)

Aircraft Owners and Pilots Association (AOPA)

Experimental Aircraft Association (EAA)

Introduction

Plane Sense is a handbook for aviation enthusiasts, especially aircraft owners and operators or those who are interested in becoming aircraft owners, who are looking for a quick reference guide on a number of general aviation topics. This handbook is published by the Federal Aviation Administration (FAA).

The FAA is the executive agency responsible for aviation oversight in the United States. The FAA's mission is to provide the safest, most efficient aerospace system in the world.

The FAA is responsible for the safety of civil aviation. The Federal Aviation Act of 1958 created the agency under the name Federal Aviation Agency. The FAA adopted its present name in 1967 when it became a part of the Department of Transportation (DOT). The FAA's major roles include:

- Regulating civil aviation to promote safety
- Encouraging and developing civil aeronautics, including new aviation technology
- Developing and operating a system of air traffic control and navigation for both civil and military aircraft
- Researching and developing the National Airspace System (NAS) and civil aeronautics
- Developing and carrying out programs to control aircraft noise and other environmental effects of civil aviation
- Regulating U.S. commercial space transportation

FAA Headquarters is located in Washington, D.C. However, the FAA is organized into eight geographical regions and the Mike Monroney Aeronautical Center located in Oklahoma City, Oklahoma. FAA regions are organized as follows:

- Alaskan (Alaska)
- Central (Iowa, Kansas, Kentucky, Missouri, Nebraska, Tennessee)
- Eastern (Connecticut, Delaware, District of Columbia, Maine, Maryland, Massachusetts, New Hampshire, New Jersey, New York, North Carolina, Pennsylvania, Rhode Island, Vermont, Virginia, West Virginia)
- Great Lakes (Illinois, Indiana, Michigan, Minnesota, North Dakota, Ohio, South Dakota, Wisconsin)
- Northwest Mountain (Colorado, Idaho, Montana, Oregon, Utah, Washington, Wyoming)
- Southern Region (Alabama, Florida, Georgia, Puerto Rico, South Carolina, Virgin Islands)
- Southwest Region (Arkansas, Louisiana, Mississippi, New Mexico, Oklahoma, Texas)
- Western-Pacific Region (Arizona, California, Hawaii, Nevada)

Within each region, you will find several Flight Standards District Offices (FSDOs). Your local FSDO is your best resource for questions about aircraft ownership, operation, maintenance, regulatory compliance, and other issues. FAA inspectors are generally assigned to a FSDO. Aviation safety inspectors (ASIs) can assist you with issues related to the operation of your aircraft, airman certification, maintenance, and other general questions.

You can find your local FSDO on the FAA website at www.faa.gov by selecting "About FAA" from the top menu bar and following the links to locate your local FSDO's contact information.

Aircraft certification-related activities are handled by the FAA Aircraft Certification Office (ACO) that serves your geographic area.

You can contact your ACO for guidance on:

- Design approval and certificate management
- U.S. production approvals
- Engineering and analysis questions
- Investigating and reporting aircraft accidents, incidents, and service difficulties
- Designated Engineering Representative (DER) oversight

You can find the nearest ACO on the FAA website at www.faa.gov by selecting "About FAA" from the top menu bar and following the links to locate the nearest ACO's contact information.

This handbook has been revised to include additional topics and updated website addresses to enable you to find relevant information more easily. You will also notice that some information is repeated in multiple chapters of *Plane Sense*, and this is designed to allow each chapter to stand alone as a reference tool to aid you in locating the information easily.

Please contact your local FSDO if you have any questions about the material in *Plane Sense* or how the information might specifically apply to your aircraft. This handbook is intended to provide general guidance for aircraft owners and operators; however, you should always ensure that the guidance provided applies to your specific aircraft and/or your specific situation.

The FAA website is continually updated. If your question is not answered in this handbook, you can access a great deal of helpful information on the FAA's website at www.faa.gov. This new edition of *Plane Sense* is color-coded to assist you in quickly finding useful information.

The chapters containing information that may be useful to all general aviation readers are coded in **blue**:

Chapter 1:	Aircraft Owner Responsibilities
Chapter 11:	Obtaining FAA Publications and Records
Appendix A:	FAA Contact Information
Appendix B:	Regulatory Guidance Index

The chapters containing information about aircraft acquisition, registration, and ownership are coded in **green**:

Chapter 2:	Buying an Aircraft
Chapter 3:	Airworthiness Certificate
Chapter 4:	Aircraft Registration
Chapter 5:	Special Flight Permits
Chapter 6:	Light Sport Aircraft

The chapters containing information on aircraft maintenance are coded in **red**:

Chapter 7:	Aircraft Maintenance
Chapter 8:	Maintenance Records
Chapter 9:	Airworthiness Directives
Chapter 10:	Service Difficulty Program

You will also note the addition of several new checklists to assist you in applying the information found in *Plane Sense*. These checklists and any relevant FAA or other forms discussed in the chapter are located at the end of that particular chapter for ease of reference. For your convenience, FAA Contact Information is at the end of this handbook.

Also, the Regulatory Guidance Index at the end of this handbook locates within each chapter the relevant regulatory guidance material, including pertinent sections of the Code of Federal Regulations (CFR), FAA Orders, and Avisory Circulars (ACs).

The FAA wishes you safe landings as you embark on your own aviation journey.

Table of Contents

1 Aircraft Owner Responsibilities

Aircraft ownership is a serious undertaking, and you should be familiar with the obligations and responsibilities of aircraft ownership before you make the decision to purchase an aircraft. Aircraft owners have a variety of responsibilities that have their foundation in the Code of Federal Regulations (CFR).

Documentation

Do you know your ARROW? Before you fly, you need to ensure that you have all of the required documentation on your aircraft. You are responsible for carrying the following documentation on your aircraft at all times:

A—Airworthiness Certificate
R—Registration Certificate
R—Radio Station License (Federal Communications Communication (FCC) Radio Station License, if required by the type of operation)
O—Operating Limitations (which may be in the form of a Federal Aviation Administration (FAA)-approved Airplane Flight Manual (AFM) and/or Pilot's Operating Handbook (POH))
W—Weight and Balance Documents

Some of these documents are addressed in this chapter, and others are covered in later chapters. *Figure 1-1* at the end if this chapter is an ARROW checklist you can use to ensure you have all of the required documents. You can expect an FAA aviation safety inspector (ASI) to ask for these documents any time he or she is inspecting your aircraft and/or assisting you with a question regarding your aircraft.

⚠ CAUTION: A radio station license is required for any international operations. You must complete an FCC Form 605, Quick-Form Application for Authorization in the Ship, Aircraft, Amateur, Restricted and Commercial Operator, and General Mobile Radio Services, available at www.fcc.gov to obtain an FCC radio station license.

Aircraft Registration

The FAA Civil Aviation Registry Aircraft Registration Branch (AFS-750) maintains registration records on individual aircraft and serves as a repository for airworthiness documents received from FAA field offices. As an aircraft owner, you are responsible for immediately notifying AFS-750 of any change of permanent mailing address, the sale or export of your aircraft, or the loss of your ability to register an aircraft in accordance with Title 14 of the Code of Federal Regulations (14 CFR) part 47, section 47.41.

Figure 1-2 at the end of this chapter is a sample FAA Change of Address Notification you can use to inform AFS-750. Aircraft registration is addressed in more detail in chapter 4.

14 CFR

14 CFR includes rules prescribed by the FAA governing all aviation activities in the United States. A wide variety of activities are regulated, such as airplane design, typical airline flights, pilot training activities, hot-air ballooning, and even model rocket launches. The rules are designed to promote safe aviation while protecting pilots, passengers, and the general public from unnecessary risk.

As an aircraft owner, you are responsible for compliance and familiarity with the applicable 14 CFR part(s) concerning the operation and maintenance of your aircraft. While the regulations cited below are not exhaustive, they are a starting point as you consider aircraft ownership. It is essential that you remember that you are responsible for complying with *all* 14 CFR parts applicable to your aircraft and aircraft operations.

- As an aircraft owner, you should be familiar with the provisions of 14 CFR Part 43, Maintenance, Preventive Maintenance, Rebuilding, and Alteration, and 14 CFR Part 91, General Operating and Flight Rules.
- If you are also a pilot, you should be familiar with the provisions of 14 CFR Part 61, Certification: Pilots, Flight Instructors, and Ground Instructors, and 14 CFR Part 67, Medical Standards and Certification.

Questions regarding 14 CFR can be addressed to your local Flight Standards District Office (FSDO). Information about obtaining copies of 14 CFR parts and FAA publications can be found in chapter 11.

Logbooks

Each aircraft has a unique set of logbooks that document historical data dating back to the manufacturing date of the aircraft. As an aircraft owner, you have a regulatory obligation to ensure that your logbooks are complete and kept up to date.

Aircraft logbooks enable the aircraft owner to keep records of the entire aircraft in chronological order including: inspections, tests, repairs, alterations, Airworthiness Directive (AD) compliance, service bulletins, and equipment additions, removals, or exchanges. Most logbooks also include sections for major alterations and altimeter/static system checks. Anyone performing maintenance on your aircraft will need complete aircraft logbooks to review the aircraft's compliance history before performing maintenance on your aircraft. Information about aircraft maintenance and aircraft maintenance records can be found in chapters 7 and 8, respectively.

Aircraft Insurance

Aircraft insurance is an important consideration for any aircraft owner. The type(s) and amount of insurance you should carry on your aircraft are influenced by several factors, and you should discuss these decisions with an insurance agent familiar with providing aviation insurance policies. Depending on the usage of your aircraft and who might be flying the aircraft, you may need Owners, Renters and/or Certified Flight Instructor (CFI) insurance policies for your aircraft. Responsible aircraft owners always carry sufficient insurance on their aircraft.

⚠ CAUTION: You should ensure that you are in compliance with any state insurance requirements relating to aircraft ownership.

Reporting Aircraft Accidents/Incidents

Aircraft owners are responsible for complying with Title 49 of the Code of Federal Regulations (49 CFR) part 830 regarding the reporting of aircraft accidents and incidents. You are required to notify the National Transportation Safety Board (NTSB) immediately of aviation accidents and certain incidents.

Accident

An accident is defined in 49 CFR part 830 as "an occurrence associated with the operation of an aircraft that takes place between the time any person boards the aircraft with the intention of flight and all such persons have disembarked, and in which any person suffers death or serious injury, or in which the aircraft receives substantial damage."

Incident

An incident is defined in 49 CFR part 830 as "an occurrence other than an accident that affects or could affect the safety of operations."

Contacting the NTSB

Contact the nearest NTSB regional office to file a report. Reports of accidents or incidents should be made to the NTSB regional office associated with the state in which the accident or incident occurred. NTSB regions are organized as follows:

- Eastern (Alabama, Connecticut, Delaware, Florida, Georgia, Kentucky, Maine, Maryland, Massachusetts, Mississippi, New Hampshire, New Jersey, New York, North Carolina, Pennsylvania, Puerto Rico, Rhode Island, South Carolina, Tennessee, Vermont, Virgin Islands, Virginia, West Virginia);
- Central (Arkansas, Colorado, Illinois, Indiana, Iowa, Kansas, Louisiana, Michigan, Minnesota, Missouri, Nebraska, New Mexico, North Dakota, Ohio, Oklahoma, South Dakota, Texas, Wisconsin);
- Western (American Samoa, Arizona, California, Guam, Hawaii, Idaho, Montana, Nevada, Oregon, Utah, Washington, Wyoming); and
- Alaska (Alaska).

Contact information for the NTSB regional offices is located at the end of this chapter in *Figure 1-3* and on the NTSB website at www.ntsb.gov. Your local FSDO can also direct you to the correct NTSB regional office in the event of an aircraft accident or incident.

For the purpose of notifying the NTSB, a phone call is sufficient initially, but a written followup is required. You will probably be directed to complete NTSB Form 6120.1, Pilot/Operator Aircraft Accident/Incident Report, which is available on the NTSB website at www.ntsb.gov, from the nearest NTSB regional office, or from your local FSDO.

Filing NTSB Form 6120.1

In accordance with 49 CFR part 830, section 830.5, you must file a report with the NTSB regional office nearest the accident or incident within 10 days after an accident for which notification is required.

Complete Form 6120.1, sign it, and send it by mail or fax to the applicable NTSB regional office. *Figure 1-4* at the end of this chapter is a sample NTSB Form 6120.1.

Aviation Safety Reporting System

The Aviation Safety Reporting System (ASRS) is an important facet of the continuing effort by government, industry, and individuals to maintain and improve aviation safety. The ASRS, which is administered by the National Aeronautics and Space Administration (NASA), collects voluntarily submitted aviation safety incident/situation reports from pilots, controllers, and others.

The ASRS acts on the information these reports contain. It identifies system deficiencies and issues alerting messages to persons in a position to correct them. The database is a public repository which serves the needs of FAA and NASA and those of other organizations worldwide engaged in research and the promotion of safe flight.

Purpose

The ASRS collects, analyzes, and responds to voluntarily submitted aviation safety incident reports in order to lessen the likelihood of aviation accidents.

ASRS data is used to:

- Identify deficiencies and discrepancies in the National Airspace (NAS) so that these can be remedied by appropriate authorities.
- Support policy formulation and planning for, and improvements to, the NAS.
- Strengthen the foundation of aviation human factors safety research. This is particularly important since it is generally conceded that over two-thirds of all aviation accidents and incidents are caused by human performance errors.

Confidentiality

Pilots, air traffic controllers, flight attendants, mechanics, ground personnel, and others involved in aviation operations submit reports to the ASRS when they are involved in, or observe, an incident or situation in which aviation safety was compromised. All submissions are voluntary.

Reports sent to the ASRS are held in strict confidence. ASRS de-identifies reports before entering them into the incident database. All personal and organizational names are removed. Dates, times, and related information that can be used to infer an identity are either generalized or eliminated.

Filing an Incident Report

An ASRS Incident Report is often referred to as a “NASA Strip.” When submitting an ASRS Incident Report, or NASA Strip, the submitter completes the form and sends it to the address indicated, and NASA returns the identification strip to the submitter as confirmation of receipt of the form.

You can obtain more information, a copy of the incident report form, or file the form electronically on the NASA website at http://asrs.arc.nasa.gov. *Figure 1-5* at the end of this chapter is a sample ASRS Incident Report.

Safety

The FAA has a number of aviation safety resources available on its website at www.faa.gov. You can access safety information by selecting "Safety" from the main menu bar.

Safety Hotline

You can call the FAA 24-Hour Safety Hotline at (800) 255-1111 or email the FAA Safety Hotline at 9-AWA-ASY-SAFETYHOTLINE@faa.gov to report:

- Maintenance improprieties
- Aircraft incidents
- Suspected unapproved parts
- Violations of 14 CFR

You can provide your contact information or file an anonymous report with the FAA Safety Hotline.

Safety Information

You can also find additional information and aviation data and statistics on the FAA website at www.faa.gov including:

- Temporary Flight Restrictions (TFRs)
- Aircraft Safety Alerts
- Safety Program Airmen Notification System
- Information for operators
- Safety alerts for operators
- CertAlerts for certificated airports
- Traffic Collision Avoidance System (TCAS) Safety Bulletin
- Aviation Safety Information Analysis and Sharing (ASIAS)
- Accident and incident data
- Aviation accident reports and statistics
- Runway incursion data and statistics
- Weather

Reporting Stolen Aircraft/Equipment

As an aircraft owner, you should be prepared to handle the theft of your aircraft or aircraft equipment. In order to manage the reporting process effectively, you should keep separate records (in a location away from the aircraft) of serial numbers for powerplants, avionics, and other installed items. Report these serial numbers at the same time the aircraft is stolen.

Law Enforcement

You should immediately report a stolen aircraft to the local law enforcement agency having jurisdiction at the site of the theft. Ask the agency to report the theft to the Federal Bureau of Investigation (FBI) National Crime Information Center, as this will initiate notifications to the appropriate government offices.

Insurance Company

After filing the appropriate reports with your local law enforcement agency, notify your insurance company or agent of the stolen aircraft, as appropriate.

Aviation Crime Prevention Institute

You should also notify the Aviation Crime Prevention Institute (ACPI) of the stolen aircraft. After you give ACPI all available information, ACPI will send notices of the theft to industry contacts, embassies, and other agencies, if applicable. You can contact the ACPI at:

Aviation Crime Prevention Institute
226 N. Nova Road
Ormond Beach, FL 32174 USA
(800) 969-5473 *toll-free*
(386) 341-7270 *outside U.S.*
(386) 615-3378 *fax*
http://www.acpi.org

Aircraft Registration Branch

If enough time has passed that the return of the aircraft is no longer expected, the owner should write to AFS-750 requesting that the registration for this aircraft be canceled. The request should fully describe the aircraft, indicate the reason for cancellation, be signed in ink by the owner, and show a title for the signer, if appropriate.

ARROW Checklist

STATUS	ITEM	DESCRIPTION
☐	A—Airworthiness certificate	FAA Form 8100-2, Standard Airworthiness Certificate, or FAA Form 8130-7, Special Airworthiness Certificate (as applicable)
☐	R—Registration certificate	FAA Form 8050-3, Certificate of Registration
☐	R—Radio station license	FCC Form 605, Quick-Form Application for Authorization in the Ship, Aircraft, Restricted and Commercial Operator, and General Mobile Radio Services, available on FCC website at www.fcc.gov (if required by the type of operation)
☐	O—Operating limitations	FAA-approved Airplane Flight Manual (AFM) and/or Pilot's Operating Handbook (POH), and/or limitations attached to FAA Form 8130-7
☐	W—Weight and balance	Documentation provided by aircraft manufacturer, maintenance and modification records

Figure 1-1. *ARROW Checklist.* You can use this checklist to ensure that you are carrying the appropriate documentation onboard your aircraft at all times.

CHANGE OF ADDRESS NOTIFICATION
(AIRCRAFT OWNER)
PRINT OR TYPE

Name of Registered Owner	Aircraft Registration Number
Joe Pilot	N 199AZ
	Manufacturer: Cessna
	Model: 172
	Serial Number: 9999

Mailing Address (if PO Box, include physical address)
123 Beechcraft Way

City Oklahoma City	State OK	Zip Code 73125
SIGNATURE (DO NOT Print or Type): Joe Pilot	Title	

SAMPLE

SIGNATURE REQUIREMENTS:
(Show appropriate title for signer)

- Individual: Owner must sign.
- Partnership: A general partner must sign.
- Corporation: A corporate officer or managing official must sign.
- Co-owner: Each Co-owner must sign.
- Government: Any authorized person may sign

AFS-750-ADCHG-1 (07/04)

(first fold)

Figure 1-2. *FAA Change of Address Notification (Aircraft Owner).* You can obtain instructions for completing an FAA Change of Address Notification on the FAA website at www.faa.gov or from your local FSDO.

NTSB Regional Offices

Eastern Region

Ashburn Regional Office
45065 Riverside Parkway
Ashburn, Virginia 20147
Phone: (571) 223-3930
Fax: (571) 223-3926

Parsippany Regional Office
2001 Route 46
Suite 310
Parsippany, New Jersey 07054
Phone: (973) 334-6420
Fax: (973) 334-6759

Atlanta Regional Office
Atlanta Federal Center
60 Forsyth Street, SW
Suite 3M25
Atlanta, Georgia 30303
Phone: (404) 562-1666
Fax: (404) 562-1674

Miami Regional Office
8405 N.W. 53rd Street
Suite B-103
Miami, Florida 33166
Phone: (305) 597-4610
Fax: (305) 597-4614

Alabama, Connecticut, Delaware, Florida, Georgia, Kentucky, Maine, Maryland, Massachusetts, Mississippi, New Hampshire, New Jersey, New York, North Carolina, Pennsylvania, Puerto Rico, Rhode Island, South Carolina, Tennessee, Vermont, Virgin Islands, Virginia, West Virginia

Central Region

Chicago Regional Office
31 West 775 North Avenue
West Chicago, Illinois 60185
Phone: (630) 377-8177
Fax: (630) 377-8172

Denver Regional Office
4760 Oakland Street
Suite 500
Denver, Colorado 80239
Phone: (303) 373-3500
Fax: (303) 373-3507

Arlington Regional Office
624 Six Flags Drive
Suite 150
Arlington, Texas 76011
Phone: (817) 652-7800
Fax: (817) 652-7803

Arkansas, Colorado, Illinois, Indiana, Iowa, Kansas, Louisiana, Michigan, Minnesota, Missouri, Nebraska, New Mexico, North Dakota, Ohio, Oklahoma, South Dakota, Texas, Wisconsin

Figure 1-3. *NTSB Regional Offices.* The updated list of NTSB Regional Offices, including office hours, is available on the NTSB website at www.ntsb.gov.

NTSB Regional Offices	
Western Region	
Seattle Regional Office 19518 Pacific Highway South Suite 201 Seattle, Washington 98188 Phone: (206) 870-2200 Fax: (206) 870-2219	**Gardena Regional Office** 1515 W. 190th Street Suite 555 Gardena, California 90248 Phone: (310) 380-5660 Fax: (310) 380-5666 **Hawaii Regional Office** Telework Location
American Samoa, Arizona, California, Guam, Hawaii, Idaho, Montana, Nevada, Oregon, Utah, Washington, Wyoming	
Alaska Region	
Anchorage Regional Office 222 West 7th Avenue Room 216, Box 11 Anchorage, Alaska 99513 Phone: (907) 271-5001 Fax: (907) 271-3007	
Alaska	

Figure 1-3. *NTSB Regional Offices* (continued).

NATIONAL TRANSPORTATION SAFETY BOARD
PILOT/OPERATOR AIRCRAFT ACCIDENT/INCIDENT REPORT
This form to be used for reporting civil and public use aircraft accidents and incidents

BASIC INFORMATION

Accident/Incident Location
Nearest City/Place: ______ State: ______
ZIP: ______ Country: ______
Latitude: ______ (dd:mm:ss N/S) Longitude: ______ (ddd:mm:ss E/W)

Date/Time
Date: ______ *mm/dd/yyyy* Local Time: ______
Time Zone: ______

Phase of Operation
☐ Standing ☐ Takeoff (incl. initial climb) ☐ Cruise ☐ Hover
☐ Taxi ☐ Climb ☐ Maneuvering ☐ Other
☐ Descent ☐ Landing ☐ Approach ☐ Unknown

Collision with Other Aircraft
☐ Midair
☐ On-ground
☐ None

Altitude of In-Flight Occurrence
______ ft MSL

AIRCRAFT INFORMATION

Manufacturer: ______
Model: ______
Serial Number: ______
Registration Number: ______ **Amateur-built:** ☐ Yes ☐ No

Max Gross Weight: ______ lbs
Weight at Time of Accident/Incident: ______ lbs
Location of Center of Gravity at Time of Accident/Incident:
______ inches from ☐ nose or ☐ datum
-or- ______ Percent Mean Aerodynamic Cord (% MAC)

Category of Aircraft
☐ Airplane
☐ Balloon
☐ Blimp/Dirigible
☐ Glider
☐ Gyrocraft
☐ Helicopter
☐ Powered lift
☐ Ultralight
☐ Unknown

Type of Airworthiness Certificate *(Check all that apply)*
Standard
☐ Normal
☐ Utility
☐ Acrobatic
☐ Transport
Special
☐ Restricted
☐ Limited
☐ Provisional
☐ Experimental
☐ Special Flight
☐ Light Sport

Number of Seats: ______
If Large Aircraft, how many seats for:
Flight Crew: ______
Cabin Crew: ______
Passengers: ______

Landing Gear ☐ Retractable
Check any additional landing gear configuration that applies:
☐ Tricycle ☐ Tailwheel
☐ Amphibian ☐ High Skid
☐ Emergency Float ☐ Skid
☐ Float ☐ Ski
☐ Hull ☐ Ski/Wheel
☐ Unknown

Type of Maintenance Program
☐ Annual
☐ Conditional (Amateur-built only)
☐ Manufacturer's Inspection Program
☐ Other Approved Inspection Program (AAIP)
☐ Continuous Airworthiness
☐ Other, specify: ______

Last Inspection Type
☐ 100 Hour ☐ Continuous Airworthiness
☐ AAIP ☐ Conditional Inspection
☐ Annual ☐ Unknown

Date Last Inspection: ______ *mm/dd/yyyy*

Airframe Total Time: ______ hrs
hours measured at *(check one)*
☐ Last Inspection ☐ Time of Accident/Incident

IFR Equipped
☐ Yes ☐ No ☐ Unknown

Stall Warning System Installed
☐ Yes ☐ No ☐ Unknown

Type of Fire Extinguishing System
☐ None
☐ Specify ______

ELT Installed
☐ Yes ☐ No

ELT Activated
☐ Yes ☐ No

ELT Aided in Locating Accident/Incident
☐ Yes ☐ No

ELT Manufacturer: ______
Model/Series: ______
Serial Number: ______
Battery Type: ______ **Battery Exp. Date:** ______

Engine Type
☐ Reciprocating ☐ Turbo Jet
☐ Turbo Shaft ☐ Turbo Fan
☐ Turbo Prop ☐ Unknown

Reciprocating Fuel System Type
☐ Carburetor
☐ Fuel Injected

Propeller
☐ Fixed Pitch
☐ Controllable Pitch
Manufacturer: ______
Model: ______

Engine	Engine Manufacturer	Engine Model/Series	Manufacturer's Serial Number	Date of Mfg. *mm/dd/yyyy*	Engine Rated Power Measured as *(check one)* ☐ Horsepower or ☐ lbs of Thrust	Total Time (hours)	Time Since Inspection (hours)	Time Since Overhaul (hours)
Eng. 1								
Eng. 2								
Eng. 3								
Eng. 4								

Figure 1-4. *NTSB Form 6120.1, Pilot/Operator Aircraft Accident/Incident Report.* You can obtain instructions for completing NTSB Form 6120.1 on the NTSB website at www.ntsb.gov or from your local FSDO.

OWNER/OPERATOR INFORMATION

Registered Aircraft Owner

Name: ____________

Fractional Ownership Aircraft: ☐ Yes ☐ No

Owner Address

City: ____________
State: ________ ZIP: ________
Country: ____________

Operator of Aircraft ☐ Same As Registered Owner

Name: ____________
Doing Business As: ____________
Air Carrier/Operator Designator (4 Character Code): ________

Operator Address ☐ Same As Registered Owner

City: ____________
State: ________ ZIP: ________
Country: ____________

Regulation Flight Conducted Under

☐ FAR 91 ☐ FAR 129 ☐ FAR 91 Special Flight ☐ Public Use (select type)
☐ FAR 103 ☐ FAR 133 ☐ Non-US, Commercial ☐ *Federal* ☐ *State* ☐ *Local*
☐ FAR 121 ☐ FAR 135 ☐ Non-US, Non-commercial ☐ Unknown
☐ FAR 125 ☐ FAR 137 ☐ Armed Forces

Revenue Sightseeing Flight

☐ Yes ☐ No

Air Medical Flight

☐ Yes ☐ No

Purpose of Flight
for FAR 91, 103, 133, 137 *(Select one)*

☐ Personal
☐ Business
☐ Executive/Corporate
☐ Other Work Use
☐ Instructional
☐ Ferry
☐ Positioning
☐ Aerial Application
☐ Aerial Observation
☐ Air Drop
☐ Air Race / Show
☐ Flight Test
☐ Public Use
☐ Unknown

Revenue Operation
for FAR 121, 125, 129, 135 *(Select one)*

☐ Scheduled or Commuter
☐ Non-Scheduled or Air Taxi

Domestic or International

☐ Domestic ☐ International

Cargo Operation

☐ Passenger/Cargo
☐ Passenger ________ How many?
☐ Cargo ________ lbs
☐ Mail

Type of Commercial Operating Certificate Held
(Check all that apply)

☐ None
☐ Flag Carrier Operating Certificate (121)
☐ Supplemental
☐ Air Cargo
☐ Foreign Air Carriers (129)
☐ Commuter Air Carrier (135)
☐ On-Demand Air Taxi (135)
☐ Large Helicopter (127)
☐ Rotorcraft External Load (133)
- or -
☐ Agricultural Aircraft (137)
☐ Other Operator of Large Aircraft

OTHER AIRCRAFT – COLLISION (If air or ground collision occurred, complete this section for *other* aircraft)

Aircraft Registration Number ____________

Manufacturer: ____________
Model: ____________

Damage to Other Aircraft

☐ Destroyed ☐ Minor
☐ Substantial ☐ None

Registered Owner of Other Aircraft

First Name: ____________ City: ____________
Middle Initial: ______ State: ________ ZIP: ________
Last Name: ____________ Country: ____________

Pilot of Other Aircraft

First Name: ____________ City: ____________
Middle Initial: ______ State: ________ ZIP: ________
Last Name: ____________ Country: ____________

MECHANICAL MALFUNCTION/FAILURE (If more space is needed, continue on separate sheet)

Was there Mechanical Malfunction/Failure? ☐ Yes ☐ No ☐ Unknown
(If yes, list the name of the part, manufacturer, part no., serial no., and describe the failure.)

Total Time/Cycles On Part

________ Hours
________ Cycles

Time Since This Part Inspected/Overhauled

________ Hours

DAMAGE TO AIRCRAFT AND OTHER PROPERTY

Aircraft Damage

☐ None ☐ Substantial
☐ Minor ☐ Destroyed

Aircraft Fire

☐ None ☐ Both Ground and In-Flight
☐ In-Flight ☐ Unknown Origin
☐ On-Ground

Aircraft Explosion

☐ None ☐ Both Ground and In-Flight
☐ In-Flight ☐ Unknown Origin
☐ On-Ground

Figure 1-4. *NTSB Form 6120.1* (page 2 of 9).

Description of Damage to Aircraft and Other Property *(use additional sheet if necessary)*

AIRPORT INFORMATION (If the accident/incident occurred on approach, takeoff or within 3 miles of an airport, complete this section)

Airport Identifier: ____________________

Airport Name: ____________________

Proximity to Airport ☐ Off Airport/Airstrip ☐ On Airport ☐ On Airstrip

Distance From Airport Center: __________ SM

Direction From Airport: __________ degrees MAG

Airport Elevation: __________ ft. MSL

Approach Segment *(Select one)*

☐ On Instrument Approach ☐ Crosswind ☐ Landing ☐ Downwind ☐ Base leg ☐ Low Approach ☐ Final ☐ Aborted Landing (after touchdown) ☐ Go Around

IFR Approach *(Check all that apply)*

☐ None ☐ ADF/NDB ☐ SDF ☐ VOR/TVOR ☐ VOR/DME ☐ TACAN ☐ PAR ☐ Sidestep ☐ ILS ☐ Localizer Only ☐ LOC-back course ☐ RNAV ☐ MLS ☐ LDA ☐ ASR ☐ Visual ☐ Contact ☐ Circling ☐ Practice ☐ GPS ☐ Loran ☐ Unknown

VFR Approach *(Check all that apply)*

☐ None ☐ Traffic Pattern ☐ Straight-In ☐ Valley/Terrain Following ☐ Go Around ☐ Full Stop ☐ Stop and Go ☐ Touch and Go ☐ Simulated Forced Landing ☐ Forced Landing ☐ Precautionary Landing ☐ Unknown

Runway Information

Runway ID: __________(L/R/C) Length: __________ft Width: __________ft

Runway/Landing Surface *(Check all that apply)*

☐ Asphalt ☐ Concrete ☐ Dirt ☐ Grass/Turf ☐ Gravel ☐ Ice ☐ Macadam ☐ Metal/Wood ☐ Snow ☐ Water ☐ Unknown

Condition of Runway/Landing Surface *(Check all that apply)*

☐ Dry ☐ Holes ☐ Ice Covered ☐ Rough ☐ Rubber Deposits ☐ Slush Covered ☐ Snow-Compacted ☐ Snow-Crusted ☐ Snow-Dry ☐ Snow-Wet ☐ Soft ☐ Vegetation ☐ Water-Calm ☐ Water-Choppy ☐ Water-Glassy ☐ Wet ☐ Unknown

FLIGHT ITINERARY INFORMATION

Last Departure Point

Airport ID: __________

City: __________

State: __________

Country: __________

Time of Departure

Time: __________

Time Zone: __________

Destination

Airport ID: __________

City: __________

State: __________

Country: __________

Type Flight Plan Filed

☐ None ☐ Company VFR ☐ Military VFR ☐ VFR ☐ VFR/IFR ☐ IFR ☐ Unknown

Activated? ☐ Yes ☐ No

Type of ATC Clearance/Service *(Check all that apply)*

☐ None ☐ VFR ☐ Special VFR ☐ IFR ☐ Special IFR ☐ VFR On Top ☐ VFR Flight Following ☐ Traffic Advisory ☐ Cruise ☐ Unknown / NA

Airspace where the accident/incident occurred *(Check all that apply)*

☐ Class A ☐ Class B ☐ Class C ☐ Class D ☐ Class E ☐ Class G ☐ Demo Area ☐ Warning Area ☐ Prohibited Area ☐ Restricted Area ☐ Military Operations Area (MOA) ☐ Airport Advisory Area ☐ Jet Training Area ☐ TRSA ☐ FAR 93 ☐ Special ☐ Air Traffic Control Area ☐ Unknown

Aircraft Load Description *(Check all that apply)*

☐ None ☐ Passengers ☐ Cargo ☐ Towing Glider ☐ Towing Banner ☐ Other External ☐ Parachutists ☐ Water ☐ Chemical/Fertilizer/Seeds ☐ Livestock ☐ Unknown

FUEL & SERVICES INFORMATION

Fuel on Board at Last Takeoff
(convert from pounds, as necessary)

__________ Gallons

Fuel Type

☐ 80/87 ☐ 100 Low Lead ☐ 100/130 ☐ 115/145 ☐ Jet A ☐ Automotive ☐ JP3 ☐ JP4 ☐ JP5 ☐ Other, specify __________

Other Services, if Any, Prior to Departure

Figure 1-4. *NTSB Form 6120.1* (page 3 of 9).

EVACUATION OF AIRCRAFT

Was an emergency evacuation of the aircraft performed? ☐ Yes ☐ No

Method of Exit – Describe how the occupants exited and how many occupants evacuated each location

WEATHER INFORMATION AT THE ACCIDENT/INCIDENT SITE

Weather Observation Facility

Facility ID: ____________

Observation Time: ____________

Time Zone: ____________

Distance from Accident Site: ____________ NM

Direction from Accident Site: ____________ degrees MAG

Source of Weather Information *(Check all that apply)*

☐ National Weather Service ☐ Flight Service Station ☐ TV/Radio ☐ Automated Report ☐ Commercial Weather Service (DUATS) ☐ Company ☐ Military ☐ Internet ☐ Unknown

Method of Briefing *(Check all that apply)*

☐ In Person ☐ Teletype ☐ Telephone/Computer ☐ Aircraft Radio ☐ TV/Radio ☐ Unknown

Briefing Type/Completeness

☐ Full ☐ Partial / Limited By Pilot ☐ Partial / Limited By Briefer ☐ Abbreviated ☐ Unknown ☐ Not Pertinent

Light Condition

☐ Dawn ☐ Day ☐ Dusk ☐ Night ☐ Dark Night ☐ Bright Night ☐ Not Reported

Visibility

____________ miles

Sky/Lowest Cloud Condition

☐ Clear ☐ Few ☐ Partial Obscuration ☐ Scattered ☐ Thin Broken ☐ Thin Overcast ☐ Unknown

Ceiling

☐ None (clear) ☐ Broken ☐ Overcast ☐ Obscured ☐ Indefinite ☐ Unknown

Restriction to Visibility *(Check all that apply)*

☐ None ☐ Blowing Dust ☐ Blowing Sand ☐ Blowing Snow ☐ Blowing Spray ☐ Dust ☐ Fog ☐ Ground Fog ☐ Haze ☐ Ice Fog ☐ Smoke ☐ Unknown

Lowest Cloud Condition Height

____________ ft AGL

Ceiling Height

____________ ft AGL

Wind Direction

☐ Indicated: ____________ degrees MAG

☐ Variable

Wind Speed

Velocity: ____________ KTS

-or-

☐ Calm ☐ Light and Variable

Wind Gusts

Velocity: ____________ KTS

☐ Gusting ☐ Not Gusting

Type of Turbulence *(Check all that apply)*

☐ None ☐ Clear Air ☐ In Clouds ☐ Vicinity of Thunderstorm

Severity of Turbulence

☐ Extreme ☐ Severe ☐ Moderate ☐ Moderate Chop ☐ Light

NOTAMs (D, L and FDC), AIRMETs, SIGMETs, PIREPs in effect at the time of the accident/incident

Temperature: ________ (C) or ________ (F)

Altimeter Setting: ________ in. HG or ________ MB

Density Altitude: ____________ ft

Dew Point: ________ (C) or ________ (F)

Icing Forecast

Amount: ☐ None ☐ Trace ☐ Light ☐ Moderate ☐ Severe

Type: ☐ Rime ☐ Clear ☐ Mixed

Icing Actual

Amount: ☐ None ☐ Trace ☐ Light ☐ Moderate ☐ Severe

Type: ☐ Rime ☐ Clear ☐ Mixed

Type of Precipitation *(Check all that apply)*

☐ None ☐ Rain ☐ Snow ☐ Hail ☐ Rain Showers ☐ Freezing Rain ☐ Snow Shower ☐ Drizzle ☐ Ice Pellets ☐ Snow Pellets ☐ Snow Grains ☐ Ice Crystals ☐ Ice Pellets Shower ☐ Freezing Drizzle

Intensity of Precipitation

☐ Light ☐ Moderate ☐ Heavy

Figure 1-4. *NTSB Form 6120.1* (page 4 of 9).

PILOT "A" INFORMATION

Pilot "A" Responsibilities at the Time of Accident/Incident

☐ Pilot ☐ Co-Pilot ☐ Student Pilot ☐ Flight Instructor ☐ Check Pilot ☐ Flight Engineer ☐ Other Flight Crew

Pilot "A" Identification

First Name: ____________ City: ____________
Middle Initial: ______ State: ______ ZIP: ______
Last Name: ____________ Country: ____________

Age at time of Accident/Incident: ______ Date of Birth: ______ *mm/dd/yyyy* Certificate Number: ____________

Degree of Injury
☐ None ☐ Fatal ☐ Minor ☐ Unknown ☐ Serious

Seat Occupied
☐ Left ☐ Front ☐ Unknown ☐ Right ☐ Rear ☐ Center ☐ Single

Seat Belt
Used ☐ Yes ☐ No
Available ☐ Yes ☐ No

Shoulder Harness
Used ☐ Yes ☐ No
Available ☐ Yes ☐ No

Pilot Certificate(s) *(Check all that apply)*
☐ None ☐ Student ☐ Recreational ☐ Commercial ☐ Flight Engineer ☐ Foreign
☐ Private ☐ Flight Instructor ☐ Sport ☐ Airline Transport ☐ U.S. Military

Principal Occupation
☐ Pilot ☐ Other ☐ Unknown

Medical Certificate
☐ None ☐ Class 3 ☐ Class 1 ☐ Driver's License (Sport Pilot only) ☐ Class 2 ☐ Unknown

Medical Certificate Validity
☐ Without limitations/waivers ☐ With limitations/waivers ☐ Unknown

Date of Last Medical
______ *mm/dd/yyyy*

Medical Certificate Limitations

Medical Certificate Waivers

Date of Last Flight Review or Equivalent, Including FAR 121/135 Checks: ______ *mm/dd/yyyy*

Flight Review Aircraft
Make: ____________
Model: ____________

Airplane Rating(s) *(Check all that apply)*
☐ None ☐ Single-Engine Land ☐ Single-Engine Sea ☐ Multiengine Land ☐ Multiengine Sea

Other Aircraft Rating(s) *(Check all that apply)*
☐ None ☐ Airship ☐ Free Balloon ☐ Glider ☐ Gyroplane ☐ Helicopter ☐ Powered Lift

Instrument Rating(s) *(Check all that apply)*
☐ None ☐ Airplane ☐ Helicopter ☐ Powered Lift

Instructor Rating(s) *(Check all that apply)*
☐ None ☐ Airplane Single-Engine ☐ Airplane Multi-Engine ☐ Gyroplane ☐ Powered Lift ☐ Instrument Airplane ☐ Instrument Helicopter ☐ Helicopter ☐ Glider ☐ Sport

Type Ratings

Student Endorsements *(Include dates)*

Flight Time *(enter appropriate number of hours in each box)*	All Aircraft	This Make & Model	Airplane Single Engine	Airplane Multiengine	Night	Instrument Actual	Instrument Simulated	Rotorcraft	Glider	Lighter Than Air
Total Time										
Pilot in Command (PIC)										
Time as Instructor										
This Make/Model	■	■	■	■				■	■	■
Last 90 Days										
Last 30 Days										
Last 24 Hours										

Figure 1-4. *NTSB Form 6120.1* (page 5 of 9).

PILOT "B" INFORMATION

Pilot "B" Responsibilities at the Time of Accident/Incident

☐ Pilot ☐ Co-Pilot ☐ Student Pilot ☐ Flight Instructor ☐ Check Pilot ☐ Flight Engineer ☐ Other Flight Crew

Pilot "B" Identification

First Name: ______________________ City: ______________________
Middle Initial: ________ State: ________ ZIP: ________
Last Name: ______________________ Country: ______________________

Age at time of Accident/Incident: ________ Date of Birth: ________ *mm/dd/yyyy* Certificate Number: ________

Degree of Injury
☐ None ☐ Fatal
☐ Minor ☐ Unknown
☐ Serious

Seat Occupied
☐ Left ☐ Front ☐ Unknown
☐ Right ☐ Rear
☐ Center ☐ Single

Seat Belt
Used ☐ Yes ☐ No
Available ☐ Yes ☐ No

Shoulder Harness
Used ☐ Yes ☐ No
Available ☐ Yes ☐ No

Pilot Certificate(s) *(Check all that apply)*
☐ None ☐ Student ☐ Recreational ☐ Commercial ☐ Flight Engineer ☐ Foreign
☐ Private ☐ Flight Instructor ☐ Sport ☐ Airline Transport ☐ U.S. Military

Principal Occupation
☐ Pilot
☐ Other
☐ Unknown

Medical Certificate
☐ None ☐ Class 3
☐ Class 1 ☐ Driver's License (Sport Pilot only)
☐ Class 2 ☐ Unknown

Medical Certificate Validity
☐ Without limitations/waivers
☐ With limitations/waivers
☐ Unknown

Date of Last Medical
________ *mm/dd/yyyy*

Medical Certificate Limitations

Medical Certificate Waivers

Date of Last Flight Review or Equivalent, Including FAR 121/135 Checks: ________ *mm/dd/yyyy*

Flight Review Aircraft
Make: ________
Model: ________

Airplane Rating(s) *(Check all that apply)*
☐ None
☐ Single-Engine Land
☐ Single-Engine Sea
☐ Multiengine Land
☐ Multiengine Sea

Other Aircraft Rating(s) *(Check all that apply)*
☐ None
☐ Airship
☐ Free Balloon
☐ Glider
☐ Gyroplane
☐ Helicopter
☐ Powered Lift

Instrument Rating(s) *(Check all that apply)*
☐ None
☐ Airplane
☐ Helicopter
☐ Powered Lift

Instructor Rating(s) *(Check all that apply)*
☐ None ☐ Instrument Airplane
☐ Airplane Single-Engine ☐ Instrument Helicopter
☐ Airplane Multi-Engine ☐ Helicopter
☐ Gyroplane ☐ Glider
☐ Powered Lift ☐ Sport

Type Ratings

Student Endorsements *(Include dates)*

Flight Time *(enter appropriate number of hours in each box)*	All Aircraft	This Make & Model	Airplane Single Engine	Airplane Multiengine	Night	Instrument		Rotorcraft	Glider	Lighter Than Air
						Actual	Simulated			
Total Time										
Pilot in Command (PIC)										
Time as Instructor										
This Make/Model	■	■	■	■				■	■	■
Last 90 Days										
Last 30 Days										
Last 24 Hours										

Figure 1-4. *NTSB Form 6120.1* (page 6 of 9).

ADDITIONAL FLIGHT CREW MEMBERS (Exclusive of cabin attendants, complete the following information)

Pilot Name and Address

First Name: ____________ City: ____________
Middle Initial: ____ State: ______ ZIP: ______
Last Name: ____________ Country: ____________

Degree of Injury
☐ None ☐ Fatal
☐ Minor ☐ Unknown
☐ Serious

Pilot Certificate(s) *(Check all that apply)*
☐ None ☐ Student ☐ Recreational ☐ Commercial ☐ Flight Engineer ☐ Foreign
☐ Private ☐ Flight Instructor ☐ Sport ☐ Airline Transport ☐ U.S. Military

Seat Occupied
☐ Left ☐ Front
☐ Right ☐ Rear
☐ Center ☐ Single
☐ Unknown

Type Rating/Endorsement for Accident/Incident Aircraft? ☐ Yes ☐ No

Total Flight Time at the Time of this Accident/Incident: ______ hrs

Pilot Name and Address

First Name: ____________ City: ____________
Middle Initial: ____ State: ______ ZIP: ______
Last Name: ____________ Country: ____________

Degree of Injury
☐ None ☐ Fatal
☐ Minor ☐ Unknown
☐ Serious

Pilot Certificate(s) *(Check all that apply)*
☐ None ☐ Student ☐ Recreational ☐ Commercial ☐ Flight Engineer ☐ Foreign
☐ Private ☐ Flight Instructor ☐ Sport ☐ Airline Transport ☐ U.S. Military

Seat Occupied
☐ Left ☐ Front
☐ Right ☐ Rear
☐ Center ☐ Single
☐ Unknown

Type Rating/Endorsement for Accident/Incident Aircraft? ☐ Yes ☐ No

Total Flight Time at the Time of this Accident/Incident: ______ hrs

Pilot Name and Address

First Name: ____________ City: ____________
Middle Initial: ____ State: ______ ZIP: ______
Last Name: ____________ Country: ____________

Degree of Injury
☐ None ☐ Fatal
☐ Minor ☐ Unknown
☐ Serious

Pilot Certificate(s) *(Check all that apply)*
☐ None ☐ Student ☐ Recreational ☐ Commercial ☐ Flight Engineer ☐ Foreign
☐ Private ☐ Flight Instructor ☐ Sport ☐ Airline Transport ☐ U.S. Military

Seat Occupied
☐ Left ☐ Front
☐ Right ☐ Rear
☐ Center ☐ Single
☐ Unknown

Type Rating/Endorsement for Accident/Incident Aircraft? ☐ Yes ☐ No

Total Flight Time at the Time of this Accident/Incident: ______ hrs

PASSENGER(S) / OTHER PERSONNEL (Include flight attendants; continue on separate sheet if necessary)

Name and Address		Seat	Crew	Non-Revenue	Revenue	Non-Occupant	FAA	Fatal	Serious Injury	Minor Injury	No Injury	Unknown
First Name: ______ Middle Initial: ____ Last Name: ______	City: ______ State: ____ ZIP: ____ Country: ______	___	☐	☐	☐	☐	☐	☐	☐	☐	☐	☐
First Name: ______ Middle Initial: ____ Last Name: ______	City: ______ State: ____ ZIP: ____ Country: ______	___	☐	☐	☐	☐	☐	☐	☐	☐	☐	☐
First Name: ______ Middle Initial: ____ Last Name: ______	City: ______ State: ____ ZIP: ____ Country: ______	___	☐	☐	☐	☐	☐	☐	☐	☐	☐	☐
First Name: ______ Middle Initial: ____ Last Name: ______	City: ______ State: ____ ZIP: ____ Country: ______	___	☐	☐	☐	☐	☐	☐	☐	☐	☐	☐
First Name: ______ Middle Initial: ____ Last Name: ______	City: ______ State: ____ ZIP: ____ Country: ______	___	☐	☐	☐	☐	☐	☐	☐	☐	☐	☐
First Name: ______ Middle Initial: ____ Last Name: ______	City: ______ State: ____ ZIP: ____ Country: ______	___	☐	☐	☐	☐	☐	☐	☐	☐	☐	☐
First Name: ______ Middle Initial: ____ Last Name: ______	City: ______ State: ____ ZIP: ____ Country: ______	___	☐	☐	☐	☐	☐	☐	☐	☐	☐	☐
First Name: ______ Middle Initial: ____ Last Name: ______	City: ______ State: ____ ZIP: ____ Country: ______	___	☐	☐	☐	☐	☐	☐	☐	☐	☐	☐

Figure 1-4. *NTSB Form 6120.1* (page 7 of 9).

NARRATIVE HISTORY OF FLIGHT (Please type or print in ink)

Describe what occurred in chronological order, including circumstances leading to and nature of accident/incident. Describe terrain and include wreckage distribution sketch if pertinent. Attach extra sheets if needed. State time and point of departure, intended destination, and services obtained.

RECOMMENDATION (How could this accident/incident have been prevented?)

Operator/Owner Safety Recommendation

Figure 1-4. *NTSB Form 6120.1* (page 8 of 9).

ADDITIONAL INFORMATION *(Please type or print in ink)*

Use this space if additional space is needed for any answers.

I HEREBY CERTIFY THAT THE ABOVE INFORMATION IS COMPLETE AND ACCURATE TO THE BEST OF MY KNOWLEDGE

Date of this Report	Signature and Name of Pilot/Operator
____________ *mm/dd/yyyy*	Signature: ____________ Type or Print Name: ____________

Signature and Name of Person Filing Report if Other than Pilot/Operator

Signature: ____________

Type or Print Name: ____________

Title: ____________

FOR NTSB USE ONLY

NTSB Accident/Incident No.	Reviewed by NTSB Regional Office	Name of Investigator	Date Report Received

Figure 1-4. *NTSB Form 6120.1* (page 9 of 9).

DO NOT REPORT AIRCRAFT ACCIDENTS AND CRIMINAL ACTIVITIES ON THIS FORM.
ACCIDENTS AND CRIMINAL ACTIVITIES ARE NOT INCLUDED IN THE ASRS PROGRAM AND SHOULD NOT BE SUBMITTED TO NASA.
ALL IDENTITIES CONTAINED IN THIS REPORT WILL BE REMOVED TO ASSURE COMPLETE REPORTER ANONYMITY.

(SPACE BELOW RESERVED FOR ASRS DATE/TIME STAMP)

IDENTIFICATION STRIP: *Please fill in all blanks to ensure return of strip.*
NO RECORD WILL BE KEPT OF YOUR IDENTITY. This section will be returned to you.

TELEPHONE NUMBERS where we may reach you for further details of this occurrence:

HOME Area ______ No ______ Hours ______
WORK Area ______ No. ______ Hours ______

NAME ______
ADDRESS/PO BOX ______
______
CITY ______ STATE ______ ZIP ______

TYPE OF EVENT/SITUATION ______
______
DATE OF OCCURRENCE ______
LOCAL TIME (24 hr. clock) ______

PLEASE FILL IN APPROPRIATE SPACES AND CHECK ALL ITEMS WHICH APPLY TO THIS EVENT OR SITUATION.

REPORTER	FLYING TIME	CERTIFICATES/RATINGS	ATC EXPERIENCE
☐ Captain ☐ First Officer ☐ pilot flying ☐ pilot not flying ☐ Other Crewmember ☐ ______	total ______ hrs. last 90 days ______ hrs. time in type ______ hrs.	☐ student ☐ private ☐ commercial ☐ ATP ☐ instrument ☐ CFI ☐ multiengine ☐ F/E ☐ ______	☐ FPL ☐ Developmental radar ______ yrs. non-radar ______ yrs. supervisory ______ yrs. military ______ yrs.

AIRSPACE	WEATHER	LIGHT/VISIBILITY	ATC/ADVISORY SERV.
☐ Class A (PCA) ☐ Class B (TCA) ☐ Class C (ARSA) ☐ Class D (Control Zone/ATA) ☐ Class E (General Controlled) ☐ Class G (Uncontrolled) ☐ Special Use Airspace ☐ airway/route ______ ☐ unknown/other ______	☐ VMC ☐ ice ☐ IMC ☐ snow ☐ mixed ☐ turbulence ☐ marginal ☐ tstorm ☐ rain ☐ windshear ☐ fog ☐ ______	☐ daylight ☐ night ☐ dawn ☐ dusk ceiling ______ feet visibility ______ miles RVR ______ feet	☐ local ☐ center ☐ ground ☐ FSS ☐ apch ☐ UNICOM ☐ dep ☐ CTAF Name of ATC Facility: ______

	AIRCRAFT 1	AIRCRAFT 2
Type of Aircraft (Make/Model)	(Your Aircraft) ______ ☐ EFIS ☐ FMS/FMC	(Other Aircraft) ______ ☐ EFIS ☐ FMS/FMC
Operator	☐ air carrier ☐ commuter ☐ military ☐ private ☐ corporate ☐ other ______	☐ air carrier ☐ commuter ☐ military ☐ private ☐ corporate ☐ other ______
Mission	☐ passenger ☐ cargo ☐ training ☐ pleasure ☐ business ☐ unk/other ______	☐ passenger ☐ cargo ☐ training ☐ pleasure ☐ business ☐ unk/other ______
Flight plan	☐ VFR ☐ IFR ☐ SVFR ☐ DVFR ☐ none ☐ unknown	☐ VFR ☐ IFR ☐ SVFR ☐ DVFR ☐ none ☐ unknown
Flight phases at time of occurrence	☐ taxi ☐ takeoff ☐ climb ☐ cruise ☐ descent ☐ approach ☐ landing ☐ missed apch/GAR ☐ other ______	☐ taxi ☐ takeoff ☐ climb ☐ cruise ☐ descent ☐ approach ☐ landing ☐ missed apch/GAR ☐ other ______
Control status	☐ visual apch ☐ controlled ☐ no radio ☐ on vector ☐ none ☐ radar advisories ☐ on SID/STAR ☐ unknown	☐ visual apch ☐ controlled ☐ no radio ☐ on vector ☐ none ☐ radar advisories ☐ on SID/STAR ☐ unknown

If more than two aircraft were involved, please describe the additional aircraft in the "Describe Event/Situation" section.

LOCATION	CONFLICTS
Altitude ______ ☐ MSL ☐ AGL	Estimated miss distance in feet: horiz ______ vert ______
Distance and radial from airport, NAVAID, or other fix ______	Was evasive action taken? ☐ Yes ☐ No
______	Was TCAS a factor? ☐ TA ☐ RA ☐ No
Nearest City/State ______	Did GPWS activate? ☐ Yes ☐ No

NASA ARC 277B (January 1994) GENERAL FORM Page 1 of 2

Figure 1-5. *ASRS Incident Report.* You can obtain an ASRS Incident Report and/or file the form online at the NASA website at http://asrs.arc.nasa.gov/.

NATIONAL AERONAUTICS AND SPACE ADMINISTRATION

NASA has established an Aviation Safety Reporting System (ASRS) to identify issues in the aviation system which need to be addressed. The program of which this system is a part is described in detail in FAA Advisory Circular 00-46D. Your assistance in informing us about such issues is essential to the success of the program. Please fill out this form as completely as possible, enclose in an sealed envelope, affix proper postage, and and send it directly to us.

The information you provide on the identity strip will be used only if NASA determines that it is necessary to contact you for further information. THIS IDENTITY STRIP WILL BE RETURNED DIRECTLY TO YOU. The return of the identity strip assures your anonymity.

AVIATION SAFETY REPORTING SYSTEM

Section 91.25 of the Federal Aviation Regulations (14 CFR 91.25) prohibits reports filed with NASA from being used for FAA enforcement purposes. This report will not be made available to the FAA for civil penalty or certificate actions for violations of the Federal Air Regulations. Your identity strip, stamped by NASA, is proof that you have submitted a report to the Aviation Safety Reporting System. We can only return the strip to you, however, if you have provided a mailing address. Equally important, we can often obtain additional useful information if our safety analysts can talk with you directly by telephone. For this reason, we have requested telephone numbers where we may reach you.

Thank you for your contribution to aviation safety.

NOTE: *AIRCRAFT ACCIDENTS SHOULD NOT BE REPORTED ON THIS FORM. SUCH EVENTS SHOULD BE FILED WITH THE NATIONAL TRANSPORTATION SAFETY BOARD AS REQUIRED BY NTSB Regulation 830.5 (49CFR830.5).*

Please fold both pages (and additional pages if required), enclose in a sealed, stamped envelope, and mail to:

NASA AVIATION SAFETY REPORTING SYSTEM
POST OFFICE BOX 189
MOFFETT FIELD, CALIFORNIA 94035-0189

DESCRIBE EVENT/SITUATION

Keeping in mind the topics shown below, discuss those which you feel are relevant and anything else you think is important. Include what you believe really caused the problem, and what can be done to prevent a recurrence, or correct the situation. (USE ADDITIONAL PAPER IF NEEDED)

CHAIN OF EVENTS
- How the problem arose
- How it was discovered
- Contributing factors
- Corrective actions

Page 2 of 2

HUMAN PERFORMANCE CONSIDERATIONS
- Perceptions, judgments, decisions
- Actions or inactions
- Factors affecting the quality of human performance

Figure 1-5. *ASRS Incident Report* (page 2 of 2).

2 Buying an Aircraft

The purchase of an aircraft represents a major commitment that should be approached carefully and cautiously, especially when buying a used aircraft. For many aircraft owners, it represents the largest single lifetime investment next to buying a home. Quite often, the purchase price of an aircraft approximates or exceeds the price of a new home.

Particularly when buying a used aircraft, it is wise to have the selected aircraft inspected by a qualified person or facility before you complete the transaction. The condition of the aircraft and the state of its maintenance records can be determined by persons familiar with the particular make and model. Pre-purchase inspections should be performed by a Federal Aviation Administration (FAA) certificated airframe and powperplant mechanic (A&P) or an approved repair station. The Buying an Aircraft Checklist (*Figure 2-1* at the end of this chapter) is a suggested list of items to consider when purchasing an aircraft.

Selecting the Aircraft

One of the most common mistakes in purchasing an aircraft is to make a decision too quickly. Take the time to analyze your requirements carefully and be realistic. Consider the typical flight loading, trip distance, and conditions of flight, then compare aircraft. If possible, rent the type of aircraft that interests you to determine how well it meets your requirements. Keep in mind that the biggest expense of owning an aircraft is not always the initial purchase price.

Where to Look

Once you have chosen the type of aircraft that will fit your needs, shop around and do some pricing. For retail and wholesale price information, check with an aviation trade association, bank, other financial institution, or Fixed Base Operator (FBO) for the latest aircraft bluebook values. There are several good publications available that advertise aircraft for sale. Your local FBO can be very helpful as you look for the right aircraft.

⚠ CAUTION: Try to keep your search for an aircraft close to home. If a problem pops up after the sale, you may not find the long-distance seller as willing to help you as someone closer to home.

Factors Affecting Resale Value

Know the major factors that affect resale value. Generally speaking they are:

- Engine hours—perhaps the most common influence on resale value. The closer an engine is to its recommended time between overhaul (TBO), the lower the value. There are many factors that affect engine health, and a high-time engine is not necessarily bad. Regular use helps keep seals and other engine components lubricated and in good shape.
- Installed equipment—such as avionics, air conditioning, deicing gear and interior equipment. The most valuable equipment is usually avionics, which can easily double the value of some older aircraft. The newer the technology, the higher the value of the aircraft.
- Airworthiness directives (ADs)—issued by the FAA for safety reasons. Once issued, owners are required to comply with the AD within the time period allotted. It is important to look at the AD history of an aircraft and ensure the logbooks show compliance with all applicable ADs. ADs are discussed in greater detail in chapter 9.
- Damage history—it may be difficult to locate a complete damage history for an aircraft. Any aircraft with a damage history should be closely scrutinized to ensure it has been properly repaired in accordance with the applicable Title 14 of the Code of Federal Regulations (14 CFR) parts and recommended practices.
- Paint/Interior—as is the case with homes, paint can be used to give "tired" aircraft a quick facelift. Check new paint jobs carefully for evidence of corrosion under the surface. Interior items should be checked for proper fit and condition.

Overhauls

Be careful of the terminology used to describe engine condition. Do not confuse a top overhaul with a major overhaul, or a major overhaul with a factory remanufactured "zero-time" engine. A top overhaul involves the repair of engine components outside of the crankcase. A major overhaul involves the complete disassembly, inspection, repair, and reassembly of an engine to specified limits. If an engine has had a top or major overhaul, the logbooks must still show the total time on the engine, if known, and its prior maintenance history. A "zero-time" engine is one that has been overhauled to factory new limits by the original manufacturer and is issued a new logbook without previous operating history.

Aircraft Records

Aircraft records maintained by the FAA are on file at the Mike Monroney Aeronautical Center in Oklahoma City, Oklahoma. Copies of aircraft records are available for review in CD format or paper. For information on ordering and costs, contact the FAA Civil Aviation Registry Aircraft Registration Branch (AFS-750). Copies of aircraft records may also be requested online. Visit www.faa.gov and select the "Aircraft Registration" link. There may be other records on file at federal, state, or local agencies that are not recorded with the FAA. AFS-750 contact information is in the FAA Contact Information appendix on pages A1–A2 of this handbook.

Make sure the following documents are available and in proper order for the aircraft:

- Airworthiness Certificate
- Engine and airframe logbooks
- Aircraft equipment list
- Weight and balance data, placards
- FAA-approved Airplane Flight Manual (AFM) and/or Pilot's Operating Handbook (POH)

⚠ CAUTION: Missing documents, pages, or entries from aircraft logbooks may cause significant problems for the purchaser and reduce the value of the aircraft.

Aircraft Title

The Federal Aviation Act requires the FAA to maintain a recording system for aircraft bills of sale, security agreements, mortgages, and other liens. This is done at AFS-750, which also processes applications for, and issues, aircraft registration certificates. The two systems are linked together because you must prove ownership in order to be entitled to register an aircraft.

"Clear title" is a term commonly used by aircraft title search companies to indicate there are no liens (e.g., chattel mortgage, security agreement, tax lien, artisan lien) in the FAA aircraft records. Title searches for the aviation public are not performed by AFS-750; however, the aircraft records contain all of the ownership and security documents that have been filed with the FAA.

AFS-750 records acceptable security instruments. In addition, some states authorize artisan liens (mechanic liens). These also need to be recorded. Be sure to check your state's statutes regarding liens.

⚠ CAUTION: Federal liens against an owner (drug, repossession, etc.) may not show up on your title search.

State law determines lien and security interests. Although there is no federal requirement to file lien or security instruments with the FAA, the parties to these transactions can file their qualifying documents with AFS-750.

You may search the aircraft records, or have this done by an attorney or aircraft title search company.

⚠ CAUTION: FAA registration cannot be used in any civil proceeding to establish proof of ownership.

There is no substitute for examining the aircraft's records to secure an ownership history and to determine if there are any outstanding liens or mortgages. This procedure should help avoid a delay in registering an aircraft.

Filing Ownership and Lien Documents

Filing ownership and lien documents constitutes formal notice to the world of the ownership and security interests recorded. A person who engages in a financial transaction involving a U.S. registered aircraft who does not have a title search performed is taking a risk. Under the law, that person will be charged with knowing what is on file with the FAA, even if he or she does not actually know. A simple title search will show the federally recorded ownership and lien status of any aircraft registered in the United States.

When a Lien Is Recorded

When a security agreement or lien document is recorded, the FAA sends an Aeronautical Center (AC) Form 8050-41, Conveyance Recordation Notice, to the secured party. This notice describes the affected aircraft (and other eligible collateral such as engines, propellers and air carrier spare parts locations). It also identifies the recorded document by its date, the parties, the FAA recording number, and date of recordation. This recordation notice is sent as a confirmation that the lien has been recorded and added to the aircraft record.

Releasing a Recorded Lien

The FAA Form 8050-41 may be used as a release if the secured party signs below the release statement and returns the form to the AFS-750. The FAA may also accept as a release a document that describes the affected collateral, specifically identifies the lien, and contains a statement releasing all lienholder rights and interest in the described collateral from the terms of the identified lien. The release document must be signed in ink by the secured party and show the signer's title, as appropriate.

A new AC Form 8050-41 may be requested by contacting AFS-750. You will need to describe the aircraft and the lien document sufficiently to identify the specific document needing release. AFS-750 contact information is in the FAA Contact Information appendix on pages A1–A2 of this handbook.

Aircraft Documents

There are numerous documents that should be reviewed and transferred when you purchase an aircraft.

Bill of Sale or Conditional Sales Contract

The bill of sale or conditional sales contract is your proof of purchase of the aircraft and will be recorded with the FAA to protect your ownership interest.

Airworthiness Certificate

The aircraft should have either FAA Form 8100-2, Standard Airworthiness Certificate, or FAA Form 8130-7, Special Airworthiness Certificate.

Maintenance Records

The previous owner of the aircraft should provide the aircraft's maintenance records containing the following information:

- The total time in service of the airframe, each engine, and each propeller;
- The current status of life-limited parts of each airframe, engine, propeller, rotor, and appliance;
- The time since last overhaul of all items installed on the aircraft that are required to be overhauled on a specified time basis;
- The identification of the current inspection status of the aircraft, including the time since the last inspection required by the inspection program under which the aircraft and its appliances are maintained;
- The current status of applicable ADs, including for each the method of compliance, the AD number, revision date, and if the AD involves recurring action, the time and date when the next action is required; and
- A copy of current major alterations to each airframe, engine, propeller, rotor, and appliance.

Manuals

Manufacturers produce owner's manuals, maintenance manuals, service letters and bulletins, and other technical data pertaining to their aircraft. These may be available from the previous owner, but are not required to be transferred to a purchaser. If the service manuals are not available from the previous owner, you can usually obtain them from the aircraft manufacturer.

Airworthiness

Two conditions must be met for a standard category aircraft to be considered airworthy:

- The aircraft conforms to its type design (type certificate). Conformity to type design is attained when the required and proper components are installed that are consistent with the drawings, specifications, and other data that are part of the type certificate. Conformity includes applicable Supplemental Type Certificate(s) (STC) and field-approval alterations.
- The aircraft is in condition for safe operation, referring to the condition of the aircraft with relation to wear and deterioration.

Maintenance

14 CFR part 91, section 91.403, places primary responsibility upon the owner for maintaining the aircraft in an airworthy condition. This includes compliance with applicable ADs. The owner is responsible for ensuring that maintenance personnel make appropriate entries in the aircraft maintenance records, indicating that the aircraft has been approved for return to service. In addition, the owner is responsible for having maintenance performed that may be required between scheduled inspections. Inoperative instruments or equipment that can be deferred under 14 CFR part 91, section 91.213(d)(2), will be placarded and maintenance recorded in accordance with 14 CFR part 43, section 43.9.

Pre-Purchase Inspection

Before buying an aircraft, you should have a mechanic you trust give the aircraft a thorough inspection and provide you with a written report of its condition. While a pre-purchase inspection need not be an annual inspection, it should include at least a differential compression check on each cylinder of the engine and any other inspections necessary to determine the condition of the aircraft. In addition to a mechanical inspection, the aircraft logbooks and other records should be carefully reviewed for such things as FAA Form 337, Report of Major Repair or Alteration, AD compliance, the status of service bulletins and letters, and aircraft/component serial numbers.

Light-Sport Aircraft

Light-sport aircraft is a growing sector of the general aviation community, specific to the United States. Several resources are available if you have questions about acquiring a light-sport aircraft. You can contact the FAA Light Sport Aviation Branch (AFS-610), your local FAA Flight Standards District Office (FSDO), or the Experimental Aircraft Association (EAA) for assistance. Chapter 6 discusses light-sport aircraft in greater detail. AFS-610 contact information is in the FAA Contact Information appendix on pages A1–A2 of this handbook.

Amateur-Built Aircraft

There are several unique considerations when purchasing an amateur-built aircraft. The prospective buyer is advised to have someone familiar with the type of aircraft check the aircraft of interest for workmanship, general construction integrity, and compliance with the applicable 14 CFR parts. You can contact your local FAA Manufacturing Inspection District Office (MIDO) or FSDO to speak with an FAA aviation safety inspector (ASI) who can explain the requirements for experimental certification.

Things to consider when buying an amateur-built aircraft:

- Examine the Special Airworthiness Certificate and its operating limitations. This certificate is used for all aircraft that fall under experimental status and states for what purpose it was issued. The operating limitations specify any operating restrictions that may apply to the aircraft.
- Check the aircraft maintenance records of the airframe, engine, propeller, and accessories. Under 14 CFR part 91, sections 91.305 and 91.319(b), all initial flight operations of experimental aircraft may be limited to an assigned flight test area. This is called Phase I. The aircraft is flown in this designated area until it is shown to be controllable throughout its normal range of speeds and all maneuvers to be executed, and that it has not displayed any hazardous operating characteristics or design features. The required flight time may vary for each type of aircraft and is covered in the operating limitations.
- After the flight time requirements are met, the owner/operator endorses the aircraft logbook with a statement certifying that the prescribed flight hours are completed and the aircraft complies with 14 CFR part 91, section 91.319(b). Phase I records are retained for the life of the aircraft.
- In Phase II, the FAA may prescribe Operating Limitations for an unlimited duration, as appropriate.
- Before taking delivery of the aircraft, make a final pre-purchase inspection. Ensure that the Special Airworthiness Certificate, Operating Limitations, Aircraft Data Plate, Weight and Balance data, Aircraft Maintenance Records, and any other required documents are with the aircraft. If the Special Airworthiness Certificate, Operating Limitations, and Aircraft Data Plate are surrendered to the FAA by the original builder, you may not be able to recertificate the aircraft because you are not the builder.
- Amateur-built aircraft require a condition inspection within the previous 12 calendar months. This inspection requirement and those who are eligible to work on the aircraft are addressed in the Operating Limitations of that particular aircraft.

Military Surplus Aircraft

Certain surplus military aircraft are not eligible for FAA certification in the STANDARD, RESTRICTED, or LIMITED classifications. The FAA, in cooperation with the Department of Defense (DOD), normally performs preliminary "screening" inspections on surplus military aircraft to determine the civil certification potential of the aircraft. For aircraft eligible for potential certification, you must "show" the FAA that your aircraft conforms to the FAA-approved type design (type certificate), and that the aircraft is in a condition for safe operation (airworthy). This means you are required to provide the technical data necessary to support this showing.

For example, certain military surplus aircraft may be eligible for certification in the RESTRICTED category and modified for special purpose operations. Military-derived RESTRICTED category aircraft may be manufactured in the United States or in a foreign country, but military surplus aircraft must be surplus of the U.S. Armed Forces. The FAA bases its certification on the operation and maintenance of the aircraft including review of the service life of the aircraft and any modifications.

When an aircraft has been modified by the military, you must either return the aircraft to the originally approved civil configuration, or obtain FAA design approval for the military modification. This is accomplished through the STC process. The STC process is also necessary for modifications to the aircraft for a special purpose operation (e.g., crop dusting). Once the FAA determines that the military surplus aircraft conforms to the FAA-approved type design, as noted in FAA Order 8130.2 (as revised), Airworthiness Certification of Aircraft and Related Products, and military records, you may apply for an airworthiness certificate.

Since no civil aircraft may be flown unless certificated, you should discuss this with an ASI at your local FSDO, who can advise you of eligible aircraft and certification procedures. An additional source for advice on amateur-built and surplus military aircraft is the EAA.

Buying an Aircraft Checklist

STATUS	ITEM	DESCRIPTION
☐	Selecting the Aircraft	Consider the location of the seller. Consider factors affecting resale value: • Engine hours • Installed equipment • ADs • Damage history • Paint/Interior Consider the condition of the engine (e.g., overhauls).
☐	Aircraft Title	Ensure the aircraft has "clear title."
☐	Aircraft Documents	Ensure the appropriate documentation is reviewed and transferred with the aircraft: • Proof of purchase (bill of sale or conditional sales contract) • Airworthiness certificate • Maintenance records
☐	Manuals	Ensure all aircraft manufacturer and other manuals are transferred with the aircraft.
☐	Maintenance	Review the maintenance records to ensure they are complete and all inspections are current.

Figure 2-1. *Buying an Aircraft Checklist.* This checklist is intended to provide a suggested list of items to consider when purchasing an aircraft. It is not an all-inclusive list, and if you have any questions, you should consult with an experienced aviation professional prior to purchasing an aircraft.

FAA FORM 8130-6, APPLICATION FOR U.S. AIRWORTHINESS CERTIFICATE

Form Approved O.M.B. No. 2120-0018
09/30/2007

U.S. Department of Transportation Federal Aviation Administration	APPLICATION FOR U.S. AIRWORTHINESS CERTIFICATE	INSTRUCTIONS - Print or type. Do not write in shaded areas; these are for FAA use only. Submit original only to an authorized FAA Representative. If additional space is required, use attachment. For special flight permits complete Sections II, VI and VII as applicable.		
I. AIRCRAFT DESIGNATION	1. REGISTRATION MARK	2. AIRCRAFT BUILDER'S NAME *(Make)*	3. AIRCRAFT MODEL DESIGNATION	4. YR. MFR.	FAA CODING
	5. AIRCRAFT SERIAL NO.	6. ENGINE BUILDER'S NAME *(Make)*	7. ENGINE MODEL DESIGNATION		
	8. NUMBER OF ENGINES	9. PROPELLER BUILDER'S NAME *(Make)*	10. PROPELLER MODEL DESIGNATION	11. AIRCRAFT IS *(Check if applicable)* IMPORT	

3 Airworthiness Certificate

An airworthiness certificate is issued by a representative of the Federal Aviation Administration (FAA) after the aircraft has been inspected, is found to meet the requirements of Title 14 of the Code of Federal Regulations (14 CFR) and is in condition for safe operation. The certificate must be displayed in the aircraft so that it is legible to passengers or crew whenever the aircraft is operated. The airworthiness certificate is transferred with the aircraft, except when it is sold to a foreign purchaser.

An airworthiness certificate is an FAA document that grants authorization to operate an aircraft in flight. The FAA provides information regarding the definition of the term "airworthy" in FAA Order 8130.2 (as revised), Airworthiness Certification of Aircraft and Related Products, chapter 1.

Classifications of Airworthiness Certificates

The FAA initially determines that your aircraft is in condition for safe operation and conforms to type design or American Society for Testing and Materials (ASTM) International standards, then issues an airworthiness certificate. There are two different classifications of airworthiness certificates: Standard Airworthiness and Special Airworthiness.

Standard Airworthiness Certificate

FAA Form 8100-2, Standard Airworthiness Certificate is the FAA's official authorization allowing for the operation of type certificated aircraft in the following categories:

- Normal
- Utility
- Acrobatic
- Commuter
- Transport
- Manned free balloons
- Special classes

A standard airworthiness certificate remains valid as long as the aircraft meets its approved type design, is in a condition for safe operation and maintenance, preventive maintenance, and alterations are performed in accordance with 14 CFR parts 21, 43, and 91.

Special Airworthiness Certificate

FAA Form 8130-7, Special Airworthiness Certificate, is an FAA authorization to operate an aircraft in U.S. airspace in one or more of categories in *Figure 3-1*.

Category	Purpose(s)	14 CFR
Primary	Aircraft flown for pleasure and personal use	Part 21, section 21.24 Part 21, section 21.184
Restricted	Aircraft with a "restricted" category type certificate, including: • Agricultural • Forest and wildlife conservation • Aerial surveying • Patrolling (pipelines, power lines) • Weather control • Aerial advertising • Other operations specified by the Administrator	Part 21, section 21.25 Part 21, section 21.185
Multiple	Multiple airworthiness certificates	Part 21, section 21.187
Limited	Aircraft with a "limited" category type certificate	Part 21, section 21.189
Light-Sport	Operate a light-sport aircraft, other than a gyroplane, kit-built, or transitioning ultralight-like vehicle	Part 21, section 21.190
Experimental	• Research and development • Showing compliance with regulations • Crew training • Exhibition • Air racing • Market surveys • Operating amateur-built aircraft • Operating kit-built aircraft • Operating light-sport aircraft • Unmanned Aircraft Systems (UAS)	Part 21, section 21.191 Part 21, section 21.193 Part 21, section 21.195
Special Flight Permit	Special-purpose flight of an aircraft that is capable of safe flight	Part 21, section 21.197
Provisional	Aircraft with a "provisional" category type certificate for special operations and operating limitations	Part 21, subpart C Part 21, subpart I Part 91, section 91.317

Figure 3-1. *Special Airworthiness Certificate Categories.*

Issuance of an Airworthiness Certificate

Only an FAA aviation safety inspector (ASI) or authorized representative of the Administrator (i.e., Designees), as defined in 14 CFR Part 183, Representatives of the Administrator, is authorized to issue an airworthiness certificate.

Your local FAA Flight Standards District Office (FSDO) processes requests for replacement airworthiness certificates. You should contact your local FSDO immediately upon discovering that you need a replacement airworthiness certificate for your aircraft.

Applying for an Airworthiness Certificate

A registered owner may apply for an airworthiness certificate by submitting FAA Form 8130-6, Application for U.S. Airworthiness Certificate, to your local FAA Manufacturing Inspection District Office (MIDO). You can find your local MIDO contact information on the FAA website at www.faa.gov. *Figure 3-2* at the end of this chapter is a sample FAA Form 8130-6. You can find instructions for completing FAA Form 8130-6 on the FAA website at www.faa.gov or in FAA Order 8130.2 (as revised). The FAA will issue the applicable certificate if the aircraft is eligible and in a condition for safe operation.

FAA Form 8100-2, Standard Airworthiness Certificate

FAA Form 8100-2, Standard Airworthiness Certificate, is issued for aircraft type certificated in the normal, utility, acrobatic, commuter, and transport categories, or for manned free balloons. The airworthiness certificate remains in effect as long as the aircraft receives the required maintenance and is properly registered in the United States. Flight safety relies, in part, on the condition of the aircraft, which may be determined on inspection by mechanics, approved repair stations, or manufacturers that meet specific requirements of 14 CFR part 43. *Figure 3-3* at the end of this chapter is a sample FAA Form 8100-2.

FAA Form 8130-7, Special Airworthiness Certificate

FAA Form 8130-7, Special Airworthiness Certificate, is issued for all aircraft certificated in other than the Standard classifications, such as Experimental, Restricted, Limited, Provisional, and Light-Sport. If you are interested in purchasing an aircraft classed as other than Standard, you should contact the local MIDO or FSDO for an explanation of airworthiness requirements and the limitations of such a certificate. The Experimental Aircraft Association (EAA) is an additional source of information on special airworthiness certificates. *Figure 3-4* at the end of this chapter is a sample FAA Form 8130-7.

⚠ CAUTION: The FAA can revoke an existing airworthiness certificate in any category (14 CFR part 21, section 21.181), if the aircraft no longer meets its approved design and/or is not in an airworthy condition.

Regulations and Policies

There are a number of regulations and policy documents that provide additional guidance on the subject of airworthiness.

Title 14 of the Code of Federal Regulations

- 14 CFR Part 21, Certification Procedures for Products and Parts
- 14 CFR Part 21, Subpart H, Airworthiness Certificates
- 14 CFR Part 45, Identification and Registration Marking
- 14 CFR Part 91, Section 91.313, Restricted category civil aircraft: Operating limitations
- 14 CFR Part 91, Subpart D, Special Flight Operations
- 14 CFR Part 91, Section 91.715, Special flight authorizations for foreign civil aircraft
- 14 CFR Part 375, Navigation of Foreign Civil Aircraft Within the United States

FAA Orders (as revised)

- FAA Order 8130.2, Airworthiness Certification of Aircraft and Related Products
- FAA Order 8900.1, Flight Standards Information Management System (FSIMS)

FAA Advisory Circulars (ACs) (as revised)

- AC 20-27, Certification and Operation of Amateur-Built Aircraft
- AC 20-139, Commercial Assistance During Construction of Amateur-Built Aircraft
- AC 21-4, Special Flight Permits for Operation of Overweight Aircraft
- AC 21-12, Application for U.S. Airworthiness Certificate, FAA Form 8130-6
- AC 45-2, Identification and Registration Marking
- AC 90-89, Amateur-Built Aircraft and Ultra-light Flight Testing Handbook

3

FAA FORM 8130-6, APPLICATION FOR U.S. AIRWORTHINESS CERTIFICATE

Form Approved O.M.B. No. 2120-0018
09/30/2007

U.S. Department of Transportation
Federal Aviation Administration

APPLICATION FOR U.S. AIRWORTHINESS CERTIFICATE

INSTRUCTIONS - Print or type. Do not write in shaded areas; these are for FAA use only. Submit original only to an authorized FAA Representative. If additional space is required, use attachment. For special flight permits complete Sections II, VI and VII as applicable.

I. AIRCRAFT DESIGNATION

1. REGISTRATION MARK	2. AIRCRAFT BUILDER'S NAME *(Make)*	3. AIRCRAFT MODEL DESIGNATION	4. YR. MFR.	FAA CODING
5. AIRCRAFT SERIAL NO.	6. ENGINE BUILDER'S NAME *(Make)*	7. ENGINE MODEL DESIGNATION		
8. NUMBER OF ENGINES	9. PROPELLER BUILDER'S NAME *(Make)*	10. PROPELLER MODEL DESIGNATION	11. AIRCRAFT IS *(Check if applicable)* IMPORT	

II. CERTIFICATION REQUESTED

APPLICATION IS HEREBY MADE FOR: *(Check applicable items)*

A 1 STANDARD AIRWORTHINESS CERTIFICATE *(Indicate Category)*: NORMAL, UTILITY, ACROBATIC, TRANSPORT, COMMUTER, BALLOON, OTHER

B SPECIAL AIRWORTHINESS CERTIFICATE *(Check appropriate items)*

- 7 PRIMARY
- 9 LIGHT-SPORT *(Indicate Class)*: AIRPLANE, POWER-PARACHUTE, WEIGHT-SHIFT-CONTROL, GLIDER, LIGHTER THAN AIR
- 2 LIMITED
- 5 PROVISIONAL *(Indicate Class)*: 1 CLASS I; 2 CLASS II
- 3 RESTRICTED *(Indicate operation(s) to be conducted)*: 1 AGRICULTURE AND PEST CONTROL; 2 AERIAL SURVEY; 3 AERIAL ADVERTISING; 4 FOREST *(Wildlife Conservation)*; 5 PATROLLING; 6 WEATHER CONTROL; 0 OTHER *(Specify)*
- 4 EXPERIMENTAL *(Indicate operation(s) to be conducted)*: 1 RESEARCH AND DEVELOPMENT; 2 AMATEUR BUILT; 3 EXHIBITION; 4 AIR RACING; 5 CREW TRAINING; 6 MARKET SURVEY; 0 TO SHOW COMPLIANCE WITH THE CFR; 7 OPERATING *(Primary Category)* KIT BUILT AIRCRAFT; 8 OPERATING LIGHT-SPORT: 8A Existing Aircraft without an airworthiness certificate & do not meet § 103.1; 8B Operating Light-Sport Kit-Built; 8C Operating light-sport previously issued special light-sport category airworthiness certificate under § 21.190
- 8 SPECIAL FLIGHT PERMIT *(Indicate operation(s) to be conducted, then complete Section VI or VII as applicable on reverse side)*: 1 FERRY FLIGHT FOR REPAIRS, ALTERATIONS, MAINTENANCE, OR STORAGE; 2 EVACUATION FROM AREA OF IMPENDING DANGER; 3 OPERATION IN EXCESS OF MAXIMUM CERTIFICATED TAKE-OFF WEIGHT; 4 DELIVERING OR EXPORTING; 5 PRODUCTION FLIGHT TESTING; 6 CUSTOMER DEMONSTRATION FLIGHTS

C 6 MULTIPLE AIRWORTHINESS CERTIFICATE *(check ABOVE "Restricted Operation" and "Standard" or "Limited" as applicable)*

III. OWNER'S CERTIFICATION

A. REGISTERED OWNER *(As shown on certificate of aircraft registration)* — IF DEALER, CHECK HERE

NAME	ADDRESS

B. AIRCRAFT CERTIFICATION BASIS *(Check applicable blocks and complete items as indicated)*

AIRCRAFT SPECIFICATION OR TYPE CERTIFICATE DATA SHEET *(Give No. and Revision No.)*	AIRWORTHINESS DIRECTIVES *(Check if all applicable AD's are compiled with and give the number of the last AD SUPPLEMENT available in the biweekly series as of the date of application)*
AIRCRAFT LISTING *(Give page number(s))*	SUPPLEMENTAL TYPE CERTIFICATE *(List number of each STC incorporated)*

C. AIRCRAFT OPERATION AND MAINTENANCE RECORDS

CHECK IF RECORDS IN COMPLIANCE WITH 14 CFR Section 91.417	TOTAL AIRFRAME HOURS	3	EXPERIMENTAL ONLY *(Enter hours flown since last certificate issued or renewed)*

D. CERTIFICATION - I hereby certify that I am the registered owner (or his agent) of the aircraft described above, that the aircraft is registered with the Federal Aviation Administration in accordance with Title 49 of the United States Code 44101 et seq. and applicable Federal Aviation Regulations, and that the aircraft has been inspected and is airworthy and eligible for the airworthiness certificate requested.

DATE OF APPLICATION	NAME AND TITLE *(Print or type)*	SIGNATURE

IV. INSPECTION AGENCY VERIFICATION

A. THE AIRCRAFT DESCRIBED ABOVE HAS BEEN INSPECTED AND FOUND AIRWORTHY BY: *(Complete the section only if 14 CFR part 21.183(d) applies.*

2	14 CFR part 121 CERTIFICATE HOLDER *(Give Certificate No.)*	3	CERTIFICATED MECHANIC *(Give Certificate No.)*	6	CERTIFICATED REPAIR STATION *(Give Certificate No.)*
5	AIRCRAFT MANUFACTURER *(Give name or firm)*				

DATE	TITLE	SIGNATURE

V. FAA REPRESENTATIVE CERTIFICATION

(Check ALL applicable block items A and B)

A. I find that the aircraft described in Section I or VII meets requirements for: 4 THE CERTIFICATE REQUESTED; AMENDMENT OR MODIFICATION OF CURRENT AIRWORTHINESS CERTIFICATE

B. Inspection for a special permit under Section VII was conducted by: FAA INSPECTOR; FAA DESIGNEE; CERTIFICATE HOLDER UNDER: 14 CFR part 65; 14 CFR part 121 OR 135; 14 CFR part 145

DATE	DISTRICT OFFICE	4	DESIGNEE'S SIGNATURE AND NO.	1	FAA INSPECTOR'S SIGNATURE

FAA Form 8130-6 (10-04) Previous Edition Dated 5/01 May be Used Until Depleted, Except for Light-Sport Aircraft   NSN: 0052-00-024-7006

Figure 3-2. *FAA Form 8130-6, Application for U.S. Airworthiness Certificate.* You can obtain instructions for completing FAA Form 8130-6 on the FAA website at www.faa.gov or from your local FSDO.

VI. PRODUCTION FLIGHT TESTING

A. MANUFACTURER	
NAME	ADDRESS

B. PRODUCTION BASIS (Check applicable item)

- [ ] PRODUCTION CERTIFICATE (Give production certificate number)
- [ ] TYPE CERTIFICATE ONLY
- [ ] APPROVED PRODUCTION INSPECTION SYSTEM

C. GIVE QUANTITY OF CERTIFICATES REQUIRED FOR OPERATING NEEDS

DATE OF APPLICATION	NAME AND TITLE (Print or Type)	SIGNATURE

VII. SPECIAL FLIGHT PERMIT PURPOSES OTHER THAN PRODUCTION FLIGHT TEST

A. DESCRIPTION OF AIRCRAFT

REGISTERED OWNER	ADDRESS
BUILDER (Make)	MODEL
SERIAL NUMBER	REGISTRATION MARK

B. DESCRIPTION OF FLIGHT CUSTOMER DEMONSTRATION FLIGHTS [ ] (Check if applicable)

FROM	TO	
VIA	DEPARTURE DATE	DURATION

C. CREW REQUIRED TO OPERATE THE AIRCRAFT AND ITS EQUIPMENT

[ ] PILOT [ ] CO-PILOT [ ] FLIGHT ENGINEER [ ] OTHER (Specify)

D. THE AIRCRAFT DOES NOT MEET THE APPLICABLE AIRWORTHINESS REQUIREMENTS AS FOLLOWS:

SAMPLE

E. THE FOLLOWING RESTRICTIONS ARE CONSIDERED NECESSARY FOR SAFE OPERATION: (Use attachment if necessary)

F. CERTIFICATION – I hereby certify that I am the registered owner (or his agent) of the aircraft described above; that the aircraft is registered with the Federal Aviation Administration in accordance with Title 49 of the United States Code 44101 et seq. and applicable Federal Aviation Regulations; and that the aircraft has been inspected and is safe for the flight described.

DATE	NAME AND TITLE (Print or Type)	SIGNATURE

VIII. AIRWORTHINESS DOCUMENTATION (FAA/DESIGNEE use only)

[ ] A. Operating Limitations and Markings in Compliance with 14 CFR Section 91.9, as applicable.	[ ] G. Statement of Conformity, FAA Form 8130-9 (Attach when required)
[ ] B. Current Operating Limitations Attached	[ ] H. Foreign Airworthiness Certification for Import Aircraft (Attach when required)
[ ] C. Data, Drawings, Photographs, etc. (Attach when required)	[ ] I. Previous Airworthiness Certificate Issued in Accordance with 14 CFR Section ________ CAR ________ (Original Attached)
[ ] D. Current Weight and Balance information Available in Aircraft	
[ ] E. Major Repair and Alteration, FAA Form 337 (Attach when required)	[ ] J. Current Airworthiness Certificate Issued in Accordance with 14 CFR Section ________ (Copy Attached)
[ ] F. This inspection Recorded in Aircraft Records	[ ] K. Light-Sport Aircraft Statement of Compliance, FAA Form 8130-15 (Attach when required)

FAA Form 8130-6 (10-04) Previous Edition Dated 5/01 May be Used Until Depleted, except for Light-Sport Aircraft NSN: 0052-00-024-7006

Figure 3-2. *FAA Form 8130-6* (page 2 of 2).

UNITED STATES OF AMERICA
DEPARTMENT OF TRANSPORTATION—FEDERAL AVIATION ADMINISTRATION
STANDARD AIRWORTHINESS CERTIFICATE

1 NATIONALITY AND REGISTRATION MARKS	2 MANUFACTURER AND MODEL	3 AIRCRAFT SERIAL NUMBER	4 CATEGORY
N2631A	PIPER PA-22-135	22-903	NORMAL

5 AUTHORITY AND BASIS FOR ISSUANCE
This airworthiness certificate is issued pursuant to the Federal Aviation Act of 1958 and certifies that, as of the date of issuance, the aircraft to which issued has been inspected and found to conform to the type certificate therefor, to be in condition for safe operation, and has been shown to meet the requirements of the applicable comprehensive and detailed airworthiness code as provided by Annex 8 to the Convention on International Civil Aviation, except as noted herein.
Exceptions:

NONE

6 TERMS AND CONDITIONS
Unless sooner surrendered, suspended, revoked, or a termination date is otherwise established by the Administrator, this airworthiness certificate is effective as long as the maintenance, preventative maintenance, and alterations are performed in accordance with Parts 21, 43, and 91 of the Federal Aviation Regulations, as appropriate, and the aircraft is registered in the United States.

DATE OF ISSUANCE	FAA REPRESENTATIVE	DESIGNATION NUMBER
08-10-95	*Marion W. Williams* MARION W. WILLIAMS	SW-FSDO-OKC

Any alteration, reproduction, or misuse of this certificate may be punishable by a fine not exceeding $1,000, or imprisonment not exceeding 3 years, or both. THIS CERTIFICATE MUST BE DISPLAYED IN THE AIRCRAFT IN ACCORDANCE WITH APPLICABLE FEDERAL AVIATION REGULATIONS.

FAA Form 8100-2 (8-82) GPO 892-804

SAMPLE

Figure 3-3. *FAA Form 8100-2, Standard Airworthiness Certificate.* The FAA issues FAA Form 8100-2, Standard Airworthiness Certificate, for aircraft type certificated in the normal, utility, acrobatic, commuter, and transport categories, or for manned free balloons.

UNITED STATES OF AMERICA
DEPARTMENT OF TRANSPORTATION - FEDERAL AVIATION ADMINISTRATION
SPECIAL AIRWORTHINESS CERTIFICATE

A	CATEGORY/DESIGNATION	EXPERIMENTAL			
	PURPOSE	OPERATING AMATEUR-BUILT AIRCRAFT			
B	MANUFAC TURER	NAME	N/A		
		ADDRESS	N/A		
C	FLIGHT	FROM	N/A		
		TO	N/A		
D	48SB			SERIAL NO.	9411
	BUILDER	STUART R. SKYE		MODEL	PITTS SIS
E	DATE OF ISSUANCE	04-01-95		EXPIRY	UNLIMITED
	OPERATING LIMITATIONS DATED 04-01-95 ARE A PART OF THIS CERTIFICATE				
	SIGNATURE OF FAA REPRESENTATIVE	Darrel A. Freeman		DESIGNATION OR OFFICE NO.	OKC-MIDO-41

Any alteration, reproduction or misuse of this certificate may be punishable by a fine not exceeding $1,000 or imprisonment not exceeding 3 years, or both. THIS CERTIFICATE MUST BE DISPLAYED IN THE AIRCRAFT IN ACCORDANCE WITH APPLICA BLE FEDERAL AVIATION REGULATIONS.

FAA Form 8130-7 (10/82) *REVERSE SIDE OF APPLICATION OF AIRWORTHINESS CERTIFICATE*

Figure 3-4. *Form 8130-7, Special Airworthiness Certificate.* The FAA issues FAA Form 8130-7, Special Airworthiness Certificate, for all aircraft certificated in other than the Standard classifications, such as Experimental, Restricted, Limited, Provisional, and Light-Sport.

REGISTRATION NOT TRANSFERABLE

UNITED STATES OF AMERICA
DEPARTMENT OF TRANSPORTATION - FEDERAL AVIATION ADMINISTRATION
CERTIFICATE OF AIRCRAFT REGISTRATION

This certificate must be in the aircraft when operated.

NATIONALITY AND REGISTRATION MARKS N48SB

AIRCRAFT SERIAL NO. 9411

MANUFACTURER AND MANUFACTURER'S DESIGNATION OF AIRCRAFT
PITTS SIS
ICAO Aircraft Address Code: 5163722

JACOBS MARK W.
520 BIPLANE LANE
TECUMSEH MI 49286

This certificate is issued for registration purposes only and is not a certif-

U.S. Department of Transportation
Federal Aviatio
Administration
Office of Aviation Sy
P.O. Box 25504
Oklahoma City, OK 73
Official Business
Penalty for Private U

4 Aircraft Registration

If you purchase an aircraft and intend to operate in the National Air Space, you must register the aircraft with the Aircraft Registration Branch (AFS-750). Aircraft may be registered under a Certificate of Aircraft Registration or Dealer's Aircraft Registration Certificate issued by AFS-750.

⚠ CAUTION: The application must be submitted in the name of the owner(s), not in the name of the bank or other mortgage holder.

Aircraft Registration Branch

You may confirm any required fees with AFS-750 prior to submitting any aircraft documents for processing. AFS-750 contact information is in the Federal Aviation Administration (FAA) Contact Information appendix on pages A1–A2 of this handbook. Visit the "Aircraft Registration" link on the FAA website at www.faa.gov for information regarding aircraft registration, recording liens, fees, importing and exporting aircraft, requesting special N-numbers, obtaining copies of aircraft records, downloading forms, an interactive aircraft registration database, etc. You may also order aircraft records by mail, fax, or telephone.

In compliance with statutory requirements, documents are processed in date-received order. You may check to see if your documents have been received by using the FAA website "Aircraft Registration" link, and selecting the "Download the Aircraft Registration Database" link to search the document index.

The FAA updates the "Aircraft Registration Inquiry" site at midnight on each federal workday. You can find new information immediately following this update. Please allow up to 20 days for processing N-number reservations and renewals and up to 30 days for all other non-priority actions.

⚠ CAUTION: The act of registration is not evidence of ownership of an aircraft in any proceeding in which ownership by a particular person is in issue. The FAA does not issue any certificate of ownership or endorse any information with respect to ownership on a Certificate of Aircraft Registration. The FAA issues a Certificate of Aircraft Registration to the person who appears to be the owner on the basis of the evidence of ownership submitted with the Aircraft Registration Application, or recorded at the FAA Aircraft Registry. Failure to properly register your aircraft may invalidate insurance, as well as have other serious consequences. You may need to follow up with AFS-750 to ensure that your aircraft registration was accomplished successfully.

⚠ CAUTION: An aircraft may not be registered in a foreign country during the period it is registered in the United States.

Eligible Registrants

An aircraft is eligible for registration in the United States if it is owned by:

- A U.S. citizen (as defined in Title 14 of the Code of Federal Regulations (14 CFR), part 47, section 47.2, a U.S. citizen can be an individual, or partnership where each individual is a U.S. citizen, or a corporation organized under the laws of the United States of which the president and at least two-thirds of the board of directors are U.S. citizens and 75 percent of the voting interest is owned or controlled by U.S. citizens)
- A resident alien
- A corporation other than one classified as a U.S. citizen, lawfully organized under the laws of the United States or of any state thereof, if the aircraft is based and used primarily in the United States
- A government entity (federal, state or local)

Registering Your Aircraft

To register an aircraft, you must send the following documentation and fee to AFS-750:

- Aeronautical Center (AC) Form 8050-1, Aircraft Registration Application,
- Evidence of ownership (such as a bill of sale), and
- The registration fee made payable to the FAA.

You must use an original AC Form 8050-1 when applying for a Certificate of Aircraft Registration. AC 8050-1 may be obtained from AFS-750 or your local FAA Flight Standards District Office (FSDO). If you use a P.O. Box as a mailing address, you must also provide your street or physical location on the application.

⚠ CAUTION: The FAA does not accept photocopies or alternate formats of AC 8050-1.

Your application for aircraft registration must include the typed or printed name of each applicant with his or her signature in the signature block.

⚠ CAUTION: An aircraft may be registered only by and in the legal name of its owner.

⚠ CAUTION: The FAA will return any applications that do not include the printed or typed name of the signer.

Figure 4-1 at the end of this chapter is an aircraft registration checklist you can use to assist you in the registration process.

Registration Number

The United States received the “N” as its nationality designator under the International Air Navigation Convention, held in 1919.

How To Form an N-Number

N-numbers consist of a series of alphanumeric characters. U.S. registration numbers may not exceed five characters in addition to the standard U.S. registration prefix letter “N.” These characters may be:

- One to five numbers (N12345),
- One to four numbers followed by one letter (N1234Z), or
- One to three numbers followed by two letters (N123AZ).

To avoid confusion with the numbers one and zero, the letters I and O are not used. Also, please note that a hyphen (-) is no longer used in U.S. registration numbers.

Other Requirements

An N-number may not begin with zero. You must precede the first zero in an N-number with any number 1 through 9. For example, N01Z is not valid.

Registration numbers N1 through N99 are strictly reserved for FAA internal use.

Special Registration Number

A special registration number is an N-number of your choice which may be reserved, if available.

Special registration numbers may be:

- Used to change the N-number currently on your aircraft.
- Assigned to a new home-built, import, or newly manufactured aircraft in preparation for registering that aircraft.
- Reserved for 1 year. Upon reservation, the FAA will mail a confirmation notice to the requester. A renewal notice will also be sent prior to the expiration date. An online reservation request program is available on the FAA website at www.faa.gov.
- Renewed annually. The renewal fee is $10 each year. An online renewal program is available on the FAA website at www.faa.gov.

Requesting a Special Registration Number
You may reserve a special N-number from the List of Available N-numbers for immediate use on a specific aircraft or for future use. This number may not exceed five characters in addition to the prefix letter "N." All five characters may be numbers (N11111) or four numbers and one suffix letter (N1000A), or one to three numbers and/or two suffix letters may be used (N100AA).

In your written request, list up to five numbers in order of preference in the event your first choice is not available, and include the fee. The fee for a Special Registration Number is $10.00. Forward your request to AFS-750.

If your request is approved, you will be notified that the number has been reserved for 1 year, and that the reservation may be extended on an annual basis for a $10 renewal fee.

Placing the Special Registration Number on Your Aircraft
When you are ready to place the number on your aircraft, you should request permission by forwarding a complete description of the aircraft to AFS-750. Permission to place the special number on your aircraft is given on AC Form 8050-64, Assignment of Special Registration Numbers. When the number is placed on your aircraft, sign and return the original AC Form 8050-64 to AFS-750 within 5 days. *Figure 4-2* at the end of this chapter is a sample AC Form 8050-64.

A duplicate AC Form 8050-64, together with your airworthiness certificate, should be presented to an aviation safety inspector (ASI) from your local FSDO within 10 days from placing the new registration number on your aircraft. The ASI will issue a revised airworthiness certificate showing the new registration number. The old registration certificate and the duplicate AC Form 8050-64 should be carried in the aircraft until the new Certificate of Registration is received, in accordance with 14 CFR part 91, section 91.203(a)(1).

Aircraft Previously Registered in the United States

If the aircraft you are purchasing was previously registered in the United States, you should immediately submit evidence of ownership, an AC Form 8050-1, Aircraft Registration Application, and the registration fee to AFS-750 upon closing. Fees required for aircraft registration may be paid by check or money order made payable to the Treasury of the United States. AFS-750 contact information is in the FAA Contact Information appendix on pages A1–A2 of this handbook.

AC Form 8050-2, Aircraft Bill of Sale, meets the FAA's requirements for evidence of ownership. An AC Form 8050-2 does not need to be notarized. AC Forms 8050-1 and 8050-2 can be obtained from the nearest FSDO, and include information and instruction sheets. *Figure 4-3* at the end of this chapter is a sample AC Form 8050-2.

⚠ CAUTION: If a conditional sales contract is the evidence of ownership, an additional fee is required for recording.

Chain of Ownership
If there is a break in the chain of ownership of the aircraft (i.e., if it is not being purchased from the last registered owner), you are required to submit conveyances to complete the chain of ownership through all intervening owners, including yourself, to AFS-750.

Replacement Certificate of Aircraft Registration
AC Form 8050-1 may also be used to report a change of address by the aircraft owner. The FAA issues a revised certificate at no charge. If the certificate is lost, destroyed, or mutilated, a replacement certificate may be obtained at the written request of the certificate holder. Send the request and fee to AFS-750.

The request should describe the aircraft by make, model, serial number, and registration number. If operation of the aircraft is necessary before receipt of the duplicate certificate, AFS-750 may, if requested, send temporary authority by fax. You should include your full address, fax number, and contact telephone number in your request AFS-750 contact information is in the FAA Contact Information appendix on pages A1–A2 of this handbook.

Aircraft Previously Registered in a Foreign Country

If you are considering the purchase of an aircraft that is currently registered in a foreign country, you should be aware that multiple issues are involved with the registration process. You should contact AFS-750 for registration assistance.

AC Form 8050-1, Aircraft Registration Application

AC Form 8050-1, Aircraft Registration Application, includes an information and instruction sheet. Submit the white and green copies to AFS-750 and keep the pink copy in the aircraft as temporary authority to operate the aircraft without registration. This temporary authority is valid until the date the applicant receives the AC Form 8050-3, Certificate of Aircraft Registration, or until the date the FAA denies the application, but in no case for more than 90 days after the date of the application. Pink copy operation is valid only inside the United States. *Figure 4-4* at the end of this chapter is a sample AC Form 8050-1.

If by 90 days the FAA has neither issued the Certificate of Aircraft Registration nor denied the application, the FAA Aircraft Registry may issue a letter of extension that serves as authority to continue to operate the aircraft without registration.

⚠ CAUTION: If you plan to operate the aircraft outside the United States within 90 days of submitting your registration documents, you should contact AFS-750 to request a temporary certificate by fax, also known as a "fly wire".

AC Form 8050-3, Certificate of Aircraft Registration

AC Form 8050-3 is issued to the person whose name is on the application. The pink copy is valid for 90 days and is legal only in the United States.

An AC Form 8050-3 should be in the aircraft before an Airworthiness Certificate can be issued. Some of the conditions under which AC Form 8050-3 becomes invalid, as described in 14 CFR part 47, section 47.41 include:

- The aircraft becomes registered under the laws of a foreign country.
- The registration of the aircraft is cancelled at the written request of the holder of the certificate.
- The aircraft is totally destroyed or scrapped.
- The holder of the certificate loses his or her U.S. citizenship or status as a resident alien without becoming a U.S. citizen.
- The ownership of the aircraft is transferred.
- Thirty days have elapsed since the death of the holder of the certificate.

When an aircraft is sold, destroyed, or scrapped, the owner must notify the FAA by filling in the back of AC Form 8050-3 and mailing it to AFS-750.

The U.S. registration and nationality marking should be removed from an aircraft before it is delivered to a purchaser who is not eligible to register it in the United States. The endorsed AC Form 8050-3 should be forwarded to AFS-750. AFS-750 contact information is in the FAA Contact Information appendix on pages A1–A2 of this handbook.

An AC Form 8050-6, Dealer's Aircraft Registration Certificate, is an alternative form of registration. It is valid only for flights within the United States by the manufacturer or dealer for flight testing or demonstration for sale. It should be removed by the dealer when the aircraft is sold.

To apply for a Dealer's Aircraft Registration Certificate, the applicant must complete AC Form 8050-5, Dealer's Aircraft Registration Certificate Application.

⚠ CAUTION: AC Form 8050-3 serves as conclusive evidence of nationality but it is not a title and is not evidence of ownership in any proceeding in which ownership is at issue.

Amateur-Built Aircraft Registration and Inspection

The FAA recommends that you apply for registration of your amateur-built aircraft 60–120 days before you finish building your aircraft, and before you submit FAA Form 8130-6 to the FAA. The FAA will not inspect your amateur-built aircraft before it has been registered or during construction of the aircraft.

The FAA or a Designated Airworthiness Representative (DAR) in your geographical area inspects your amateur-built aircraft for general airworthiness only after you have made an application for an airworthiness certificate.

The FAA does not charge a fee to the public for inspecting amateur-built aircraft. However, FAA workload may delay inspection of your aircraft. For this reason, the FAA staff is augmented by the use of DARs who may charge a fee for their services (14 CFR Part 183, Section 183.33(b), Designated Airworthiness Representative).

You may locate a DAR in your geographical area by reviewing the online DAR Directory. Manufacturing DARs are listed by state in the first half of the directory, and Maintenance DARs are listed by state in the second half. A DAR who has authority to inspect and certify amateur-built aircraft has the

DAR Function Code "46" under his or her name. (Also see FAA Order 8100.8 (as revised), Designee Management Handbook, for designee program details.)

Light-Sport Aircraft Registration

If you purchased a newly manufactured light-sport aircraft that is to be certificated as:

- An experimental light-sport aircraft under 14 CFR, part 21, section 21.191(i)(2); or
- A special light-sport aircraft under 14 CFR part 21, section 21.190;

Then you must provide the following documentation to AFS-750:

- AC Form 8050-88 (as revised), Light-Sport Aircraft Manufacturer's Affidavit, completed by the light-sport aircraft manufacturer, unless previously submitted to AFS-750 by the manufacturer;
- Evidence of ownership from the manufacturer for the aircraft;
- AC Form 8050-1, Aircraft Registration Application; and
- Registration fee.

The FAA Light Sport Aviation Branch (AFS-610) or your local FSDO can assist you with questions about the registration of light-sport aircraft. AFS-610 contact information is in the FAA Contact Information appendix on pages A1–A2 of this handbook.

State Registration Requirements

Aircraft owners should remember that state registration of aircraft is required in many states. You should check with your state government to ensure that you have met any applicable state registration requirements for your aircraft.

Additional Information

14 CFR part 47 specifies the requirements for registering an aircraft. For information concerning 14 CFR part 47 or any topics not discussed in this chapter, please contact AFS-750. AFS-750 contact information is in the FAA Contact Information appendix on pages A1–A2 of this handbook.

Aircraft Registration Checklist

Figure 4-1. *Aircraft Registration Checklist.* You can use this checklist to assist you with the aircraft

STATUS	ITEM	DESCRIPTION
☐	Certificate of airworthiness	Confirm the aircraft has a valid certificate of airworthiness or special certificate of airworthiness, if applicable.
☐	Eligible registrant	In accordance with 14 CFR part 47: U.S. citizen, resident alien, corporation or government entity
☐	Aircraft Registration Application	Original AC Form 8050-1, Aircraft Registration Application
☐	Evidence of ownership	AC Form 8050-2, Bill of Sale or conditional sales contract
☐	Fee	Check with AFS-750 to determine applicable fee (i.e., conditional sales contract as evidence of ownership triggers additional fee).
☐	Registration number	Confirm registration number on aircraft.
☐	Special registration number (if applicable)	Complete FAA Form 8050-64, Assignment of Special Registration Number, when you are ready to use the new registration number.
☐	Chain of ownership	Ensure that chain of ownership is uninterrupted. If chain of ownership is interrupted, you must submit conveyances completing the chain of ownership to AFS-750.
☐	State registration requirements	Check with your state aviation authority to confirm any state requirements.
☐	Light-sport aircraft (if applicable)	You must send the following documents to AFS-750: AC Form 8050-88A, Light-Sport Aircraft Manufacturer's Affidavit; evidence of ownership from the manufacturer for the aircraft; AC Form 8050-1, Aircraft Registration Application; and the registration fee.

U.S. Department of Transportation

Federal Aviation Administration

ASSIGNMENT OF SPECIAL REGISTRATION NUMBERS

Special Registration Number	N401RZ
Aircraft Make and Model	CIRRUS DESIGN CORP SR22
Present Registration Number	N402TS
Serial Number	2917
Issue Date:	JUN 13, 2008

ICAO AIRCRAFT ADDRESS CODE FOR N401RZ - 50999999

RANDALL Z. BLACKHAWK
150 CESSNA ROAD
OKLAHOMA CITY, OK 73125

This is your authority to change the United States registration number on the above described aircraft to the special registration number shown.

Carry duplicate of this form in the aircraft together with the old registration certificate as interim authority to operate the aircraft pending receipt of revised certificate of registration. Obtain a revised certificate of airworthiness from your nearest Flight Standards District Office.

The latest FAA Form 8130-6, Application For Airworthiness on file is dated:
JAN 30, 2008

The airworthiness classification and category:
STANDARD

INSTRUCTIONS:
SIGN AND RETURN THE ORIGINAL of this form to the Civil Aviation Registry, AFS-750, within 5 days after the special registration number is placed on the aircraft. A revised certificate will then be issued.

The authority to use the special number expires: JUN 13, 2009

CERTIFICATION: I certify that the special registration number was placed on the aircraft described above.

Signature of Owner:

Title of Owner:

Date Placed on Aircraft:

RETURN FORM TO:

Civil Aviation Registry, AFS-750
P.O. Box 25504
Oklahoma City, Oklahoma 73125-0504

Figure 4-2. *AC Form 8050-64, Assignment of Special Registration Numbers.* The FAA issues AC Form 8050-64 to give you permission to place your reserved special registration number on your aircraft. You should place the special registration number on your aircraft, and then notify the FAA in accordance with the instructions provided.

4

FORM APPROVED
OMB NO. 2120-0042
08/31/2008

UNITED STATES OF AMERICA
U.S. DEPARTMENT OF TRANSPORTATION FEDERAL AVIATION ADMINISTRATION

AIRCRAFT BILL OF SALE

FOR AND IN CONSIDERATION OF $ THE UNDERSIGNED OWNER(S) OF THE FULL LEGAL AND BENEFICIAL TITLE OF THE AIRCRAFT DESCRIBED AS FOLLOWS:

UNITED STATES REGISTRATION NUMBER **N** 103AZ

AIRCRAFT MANUFACTURER & MODEL CESSNA 172

AIRCRAFT SERIAL No. 54320

DOES THIS 5TH DAY OF JUNE , 2006 HEREBY SELL, GRANT, TRANSFER AND DELIVER ALL RIGHTS, TITLE, AND INTERESTS IN AND TO SUCH AIRCRAFT UNTO:

Do Not Write In This Block
FOR FAA USE ONLY

PURCHASER

NAME AND ADDRESS
(IF INDIVIDUAL(S), GIVE LAST NAME, FIRST NAME, AND MIDDLE INITIAL.)

FRED WINGTIP
44 ECLIPSE STREET
OKLAHOMA CITY, OK 73125

DEALER CERTIFICATE NUMBER

AND TO EXECUTORS, ADMINISTRATORS, AND ASSIGNS TO HAVE AND TO HOLD SINGULARLY THE SAID AIRCRAFT FOREVER AND WARRANTS THE TITLE THEREOF:

IN TESTIMONY WHEREOF HAVE SET HAND AND SEAL THIS DAY OF

SELLER

NAME(S) OF SELLER (TYPED OR PRINTED)	SIGNATURE(S) (IN INK) (IF EXECUTED FOR CO-OWNERSHIP, ALL MUST SIGN.)	TITLE (TYPED OR PRINTED)
JANE FLYER	Jane Flyer	OWNER

ACKNOWLEDGMENT (NOT REQUIRED FOR PURPOSES OF FAA RECORDING: HOWEVER, MAY BE REQUIRED BY LOCAL LAW FOR VALIDITY OF THE INSTRUMENT.)

ORIGINAL: TO FAA:

AC Form 8050-2 (9/92) (NSN 0052-00-629-0003) Supersedes Previous Edition

SAMPLE

Figure 4-3. *AC Form 8050-2, Aircraft Bill of Sale.* You can download the form and obtain instructions for completing FAA Form 8050-2 on the FAA website at www.faa.gov or from your local FSDO.

FORM APPROVED
OMB No. 2120-0042

UNITED STATES OF AMERICA DEPARTMENT OF TRANSPORTATION
FEDERAL AVIATION ADMINISTRATION-MIKE MONRONEY AERONAUTICAL CENTER
AIRCRAFT REGISTRATION APPLICATION

CERT. ISSUE DATE

UNITED STATES REGISTRATION NUMBER N 142AZ

AIRCRAFT MANUFACTURER & MODEL
Beechcraft A200

AIRCRAFT SERIAL No.
BO-12

FOR FAA USE ONLY

TYPE OF REGISTRATION (Check one box)

☒ 1. Individual ☐ 2. Partnership ☐ 3. Corporation ☐ 4. Co-owner ☐ 5. Gov't. ☐ 8. Non-Citizen Corporation

NAME OF APPLICANT (Person(s) shown on evidence of ownership. If individual, give last name, first name, and middle initial.)

GLENN NAVIGATOR

TELEPHONE NUMBER: (405) 555-8118

ADDRESS (Permanent mailing address for first applicant listed.) (If P.O. BOX is used, physical address must also be shown.)

Number and street: 14 SKYCATCHER BLVD.

Rural Route: P.O. Box:

CITY	STATE	ZIP CODE
OKLAHOMA CITY	OK	73125

SAMPLE

☐ **CHECK HERE IF YOU ARE ONLY REPORTING A CHANGE OF ADDRESS**

ATTENTION! Read the following statement before signing this application.
This portion MUST be completed.

A false or dishonest answer to any question in this application may be grounds for punishment by fine and / or imprisonment (U.S. Code, Title 18, Sec. 1001).

CERTIFICATION

I/WE CERTIFY:

(1) That the above aircraft is owned by the undersigned applicant, who is a citizen (including corporations) of the United States.

(For voting trust, give name of trustee: ______), or

CHECK ONE AS APPROPRIATE:

a. ☐ A resident alien, with alien registration (Form 1-151 or Form 1-551) No. ______

b. ☐ A non-citizen corporation organized and doing business under the laws of (state) ______ and said aircraft is based and primarily used in the United States. Records or flight hours are available for inspection at ______

(2) That the aircraft is not registered under the laws of any foreign country; and

(3) That legal evidence of ownership is attached or has been filed with the Federal Aviation Administration.

NOTE: If executed for co-ownership all applicants must sign. Use reverse side if necessary.

TYPE OR PRINT NAME BELOW SIGNATURE

EACH PART OF THIS APPLICATION MUST BE SIGNED IN INK.

SIGNATURE	TITLE	DATE
Glenn Navigator GLENN NAVIGATOR	OWNER	04-28-06
SIGNATURE	TITLE	DATE
SIGNATURE	TITLE	DATE

NOTE Pending receipt of the Certificate of Aircraft Registration, the aircraft may be operated for a period not in excess of 90 days, during which time the PINK copy of this application must be carried in the aircraft.

AC Form 8050-1 (5/03) (0052-00-628-9007)

Figure 4-4. *AC Form 8050-1, Aircraft Registration Application.* You must use an original AC Form 8050-1 which can be obtained from AFS-750 or your local FSDO. You can obtain instructions for completing AC Form 8050-1 on the FAA website at www.faa.gov or from your local FSDO. (Be sure to print your name below your signature or your application will be rejected.)

REGISTRATION NOT TRANSFERABLE

UNITED STATES OF AMERICA
DEPARTMENT OF TRANSPORTATION - FEDERAL AVIATION ADMINISTRATION
CERTIFICATE OF AIRCRAFT REGISTRATION

This certificate must be in the aircraft when operated.

NATIONALITY AND REGISTRATION MARKS N505DH

AIRCRAFT SERIAL NO. 8806

MANUFACTURER AND MANUFACTURER'S DESIGNATION OF AIRCRAFT
PITTS SIS
ICAO Aircraft Address Code: 5199999

ISSUED TO
LINBERGH, DANIEL E.
800 GATEWAY ROAD
OKLAHOMA CITY, OK 73125

This certificate is issued for registration purposes only and is not a certificate of title. The Federal Aviation Administration does not determine rights of ownership as between private persons.

It is certified that the above described aircraft has been entered on the register of the Federal Aviation Administration, United States of America, in accordance with the Convention on International Civil Aviation dated December 7, 1944, and with the Federal Aviation Act of 1958, and regulations issued thereunder.

DATE OF ISSUE
JUNE 3,1995

David Hinson
ADMINISTRATOR

U.S. Department of Transportation
Federal Aviation Administration

AC Form 8050-3(11/93) Supersedes previous editions

Figure 4-5. *AC Form 8050-3, Certificate of Aircraft Registration.* The FAA issues AC Form 8050-3 to evidence registration of your aircraft.

5 Special Flight Permits

A special flight permit can be issued to any U.S. registered aircraft that may not currently meet applicable airworthiness requirements but is capable of safe flight. Before the permit is issued, a Federal Aviation Administration (FAA) aviation safety inspector (ASI) may inspect the aircraft or require it to be inspected by an FAA-certificated airframe and powerplant (A&P) mechanic or repair station to determine its safety for the intended flight. The inspection is then recorded in the aircraft records. This type of special flight permit is often referred to as a "ferry permit" because it allows the aircraft to be ferried to a location for maintenance.

In the case of general aviation flights (e.g., flights conducted by operators other than Title 14 of the Code of Federal Regulations (14 CFR) part 121 or part 135 certificate holders), special flight permits are issued by the FAA Flight Standards District Office (FSDO)/International Field Office (IFO) having jurisdiction over the geographical area in which the flight is to originate.

Circumstances Warranting a Special Flight Permit

A special flight permit is issued to allow the aircraft to be flown to a base where repairs, alterations, or maintenance can be performed; for delivering or exporting the aircraft; or for evacuating an aircraft from an area of impending danger. It may also be issued to allow the operation of an overweight aircraft for flight beyond its normal range over water or land areas where adequate landing facilities are not available.

The following list, which is not all-inclusive, sets forth the most common requests for special flight permits:

- Flying the aircraft to a base where repairs, alterations, or maintenance are to be performed, or to a point of salvage
- Flying an aircraft whose annual inspection has expired to a base where an annual inspection can be accomplished
- Flying an amateur-built aircraft whose condition inspection has expired to a base where the condition inspection can be accomplished
- Delivering or exporting the aircraft
- Production flight testing of new production aircraft
- Evacuating aircraft from areas of impending danger
- Conducting customer demonstration flights in new production aircraft that have satisfactorily completed production flight tests
- Operating an aircraft at a weight in excess of its maximum certificated takeoff weight

⚠ CAUTION: If an Airworthiness Directive (AD) requires compliance before further flight and does not have a provision for issuance of a special flight permit, the operation of the aircraft to which it applies would not be appropriate, and a special flight permit will not be issued.

Foreign-Registered Civil Aircraft

A special flight authorization allows a foreign-registered civil aircraft that does not have the equivalent of a U.S. standard airworthiness certificate to be operated within the United States.

A civil aircraft registered in a country that is a member of the International Civil Aviation Organzation (ICAO) only needs a special flight authorization issued by the FAA. A civil aircraft registered in a country that is not a member of ICAO must have both an authorization from the United States Department of Transportation (DOT) and a special flight authorization issued by the FAA.

Obtaining a Special Flight Authorization

To obtain a special flight authorization, you must apply by letter or facsimile. The application and issuance procedures are provided in FAA Order 8130.2 (as revised), Airworthiness Certification of Aircraft and Related Products, Chapter 7, Special Flight Authorizations for Non-U.S.-Registered Civil Aircraft. You can obtain assistance and the necessary forms for issuance of a special flight authorization from the local FSDO or IFO.

Application for Airworthiness Certificate

A special flight permit is an FAA Form 8130-7, Special Airworthiness Certificate, issued pursuant to 14 CFR part 21, section 21.197, for an aircraft that may not currently meet applicable airworthiness requirements but is safe for a specific flight.

⚠ CAUTION: A special flight permit is not an authorization to deviate from the requirements of 14 CFR part 91.

An applicant for a special flight permit must submit FAA Form 8130-6, Application for U.S. Airworthiness Certificate, including a statement indicating:

- Purpose of the flight;
- Proposed itinerary;
- Essential crew required to operate the aircraft;
- The ways, if any, in which the aircraft does not comply with the applicable airworthiness requirements; and
- Any other information requested by the Administrator, considered necessary for the purpose of prescribing operating limitations.

You should fax the completed form to the FSDO closest to the location where the flight will originate. You can locate contact information for the FSDO on the FAA website at www.faa.gov. *Figure 5-1* at the end of this chapter is a sample FAA Form 8130-6. Some FSDOs may ask you to include additional information necessary for the purpose of prescribing operating limitations when you submit FAA Form 8130-6. Examples of additional information might include:

- A current copy of the Airworthiness Certificate.
- A current copy of the Certificate of Aircraft Registration.
- A current copy of the front page of the aircraft and engine(s) logbooks, with all entries completed, (i.e., aircraft, engine(s), propeller(s), manufacturer, model, serial number).
- A current copy of the Aircraft/Engine/Propeller/Appliance AD compliance status.
- A copy of the last Aircraft Log Book entry, stating that the aircraft has been inspected and is in a safe condition to fly/ferry, that the aircraft is in compliance with all applicable ADs, and/or a listing of the ADs with which the aircraft is not in compliance. The entry must be signed by an A&P mechanic or 14 CFR part 145 repair station.

The Administrator may also want to inspect the aircraft in question before approving or issuing a ferry flight permit.

You may be required to make appropriate inspections or tests necessary for safety. (This means an A&P mechanic or 14 CFR part 145 repair station will need to inspect the aircraft prior to flight.)

You may request that the local FSDO transmit the ferry permit via facsimile if the request is time-sensitive. *Figure 5-2* at the end of this chapter is a sample FAA Form 8130-7.

The aircraft operator must display, in the aircraft, the current airworthiness certificate and the special flight (ferry) permit along with its operating limitations.

For additional information, please refer to 14 CFR part 21, section 21.197, or your local FSDO.

FAA FORM 8130-6, APPLICATION FOR U.S. AIRWORTHINESS CERTIFICATE

Form Approved O.M.B. No. 2120-0018
09/30/2007

U.S. Department of Transportation
Federal Aviation Administration

APPLICATION FOR U.S. AIRWORTHINESS CERTIFICATE

INSTRUCTIONS - Print or type. Do not write in shaded areas; these are for FAA use only. Submit original only to an authorized FAA Representative. If additional space is required, use attachment. For special flight permits complete Sections II, VI and VII as applicable.

I. AIRCRAFT DESIGNATION

1. REGISTRATION MARK	2. AIRCRAFT BUILDER'S NAME *(Make)*	3. AIRCRAFT MODEL DESIGNATION	4. YR. MFR.	FAA CODING
5. AIRCRAFT SERIAL NO.	6. ENGINE BUILDER'S NAME *(Make)*	7. ENGINE MODEL DESIGNATION		
8. NUMBER OF ENGINES	9. PROPELLER BUILDER'S NAME *(Make)*	10. PROPELLER MODEL DESIGNATION	11. AIRCRAFT IS *(Check if applicable)*	IMPORT

II. CERTIFICATION REQUESTED

APPLICATION IS HEREBY MADE FOR: *(Check applicable items)*

- A 1 STANDARD AIRWORTHINESS CERTIFICATE *(Indicate Category)*: NORMAL / UTILITY / ACROBATIC / TRANSPORT / COMMUTER / BALLOON / OTHER
- B X SPECIAL AIRWORTHINESS CERTIFICATE *(Check appropriate items)*
  - 7 PRIMARY
  - 9 LIGHT-SPORT *(Indicate Class)*: AIRPLANE / POWER-PARACHUTE / WEIGHT-SHIFT-CONTROL / GLIDER / LIGHTER THAN AIR
  - 2 LIMITED
  - 5 PROVISIONAL *(Indicate Class)*: 1 CLASS I; 2 CLASS II
  - 3 RESTRICTED *(Indicate operation(s) to be conducted)*: 1 AGRICULTURE AND PEST CONTROL; 2 AERIAL SURVEY; 3 AERIAL ADVERTISING; 4 FOREST *(Wildlife Conservation)*; 5 PATROLLING; 6 WEATHER CONTROL; 0 OTHER *(Specify)*
  - 4 EXPERIMENTAL *(Indicate operation(s) to be conducted)*: 1 RESEARCH AND DEVELOPMENT; 2 AMATEUR BUILT; 3 EXHIBITION; 4 AIR RACING; 5 CREW TRAINING; 6 MARKET SURVEY; 0 TO SHOW COMPLIANCE WITH THE CFR; 7 OPERATING *(Primary Category)* KIT BUILT AIRCRAFT; 8 OPERATING LIGHT-SPORT: 8A Existing Aircraft without an airworthiness certificate & do not meet § 103.1; 8B Operating Light-Sport Kit-Built; 8C Operating light-sport previously issued special light-sport category airworthiness certificate under § 21.190
  - 8 X SPECIAL FLIGHT PERMIT *(Indicate operation(s) to be conducted, then complete Section VI or VII as applicable on reverse side)*: 1 X FERRY FLIGHT FOR REPAIRS, ALTERATIONS, MAINTENANCE, OR STORAGE; 2 EVACUATION FROM AREA OF IMPENDING DANGER; 3 OPERATION IN EXCESS OF MAXIMUM CERTIFICATED TAKE-OFF WEIGHT; 4 DELIVERING OR EXPORTING; 5 PRODUCTION FLIGHT TESTING; 6 CUSTOMER DEMONSTRATION FLIGHTS
- C 6 MULTIPLE AIRWORTHINESS CERTIFICATE *(check ABOVE "Restricted Operation" and "Standard" or "Limited" as applicable)*

SAMPLE

III. OWNER'S CERTIFICATION

A. REGISTERED OWNER *(As shown on certificate of aircraft registration)* — IF DEALER, CHECK HERE

NAME | ADDRESS

B. AIRCRAFT CERTIFICATION BASIS *(Check applicable blocks and complete items as indicated)*

- AIRCRAFT SPECIFICATION OR TYPE CERTIFICATE DATA SHEET *(Give No. and Revision No.)*
- AIRWORTHINESS DIRECTIVES *(Check if all applicable AD's are compiled with and give the number of the last AD SUPPLEMENT available in the biweekly series as of the date of application)*
- AIRCRAFT LISTING *(Give page number(s))*
- SUPPLEMENTAL TYPE CERTIFICATE *(List number of each STC incorporated)*

C. AIRCRAFT OPERATION AND MAINTENANCE RECORDS

- CHECK IF RECORDS IN COMPLIANCE WITH 14 CFR Section 91.417
- TOTAL AIRFRAME HOURS
- 3 EXPERIMENTAL ONLY *(Enter hours flown since last certificate issued or renewed)*

D. CERTIFICATION - I hereby certify that I am the registered owner (or his agent) of the aircraft described above, that the aircraft is registered with the Federal Aviation Administration in accordance with Title 49 of the United States Code 44101 et seq. and applicable Federal Aviation Regulations, and that the aircraft has been inspected and is airworthy and eligible for the airworthiness certificate requested.

DATE OF APPLICATION | NAME AND TITLE *(Print or type)* | SIGNATURE

IV. INSPECTION AGENCY VERIFICATION

A. THE AIRCRAFT DESCRIBED ABOVE HAS BEEN INSPECTED AND FOUND AIRWORTHY BY: *(Complete the section only if 14 CFR part 21.183(d) applies.*

- 2 14 CFR part 121 CERTIFICATE HOLDER *(Give Certificate No.)*
- 3 CERTIFICATED MECHANIC *(Give Certificate No.)*
- 6 CERTIFICATED REPAIR STATION *(Give Certificate No.)*
- 5 AIRCRAFT MANUFACTURER *(Give name or firm)*

DATE | TITLE | SIGNATURE

V. FAA REPRESENTATIVE CERTIFICATION

(Check ALL applicable block items A and B)

A. I find that the aircraft described in Section I or VII meets requirements for — THE CERTIFICATE REQUESTED; 4 AMENDMENT OR MODIFICATION OF CURRENT AIRWORTHINESS CERTIFICATE

B. Inspection for a special permit under Section VII was conducted by: FAA INSPECTOR; X CERTIFICATE HOLDER UNDER; FAA DESIGNEE; X 14 CFR part 65; 14 CFR part 121 OR 135; 14 CFR part 145

DATE	DISTRICT OFFICE	4 DESIGNEE'S SIGNATURE AND NO.	1 FAA INSPECTOR'S SIGNATURE
2-26-2003	SW 15		Joe Pilot

FAA Form 8130-6 (10-04) Previous Edition Dated 5/01 May be Used Until Depleted, Except for Light-Sport Aircraft NSN: 0052-00-024-7006

Figure 5-1. *FAA Form 8130-6, Application for U.S. Airworthiness Certificate.* You can obtain instructions for completing FAA Form 8130-6 on the FAA website at www.faa.gov or from your local FSDO.

VI. PRODUCTION FLIGHT TESTING

A. MANUFACTURER	
NAME	ADDRESS
B. PRODUCTION BASIS (Check applicable item)	
☐ PRODUCTION CERTIFICATE (Give production certificate number)	
☐ TYPE CERTIFICATE ONLY	
☐ APPROVED PRODUCTION INSPECTION SYSTEM	
C. GIVE QUANTITY OF CERTIFICATES REQUIRED FOR OPERATING NEEDS	

DATE OF APPLICATION	NAME AND TITLE (Print or Type)	SIGNATURE

VII. SPECIAL FLIGHT PERMIT PURPOSES OTHER THAN PRODUCTION FLIGHT TEST

A. DESCRIPTION OF AIRCRAFT

REGISTERED OWNER JANE A. AVIATOR	ADDRESS 1012 CIRRUS AVE., SHAWNEE, OKLAHOMA 74852
BUILDER (Make) CESSNA	MODEL C-182L
SERIAL NUMBER 182-500000	REGISTRATION MARK N122A

B. DESCRIPTION OF FLIGHT — CUSTOMER DEMONSTRATION FLIGHTS ☐ (Check if applicable)

FROM SHAWNEE, OKLAHOMA	TO DOWNTOWN AIRPARK, OKLAHOMA CITY, OKLAHOMA	
VIA DIRECT	DEPARTURE DATE 2-26-2008	DURATION 10 DAYS

C. CREW REQUIRED TO OPERATE THE AIRCRAFT AND ITS EQUIPMENT

☐ PILOT ☐ CO-PILOT ☐ FLIGHT ENGINEER ☐ OTHER (Specify)

D. THE AIRCRAFT DOES NOT MEET THE APPLICABLE AIRWORTHINESS REQUIREMENTS AS FOLLOWS:

ANNUAL INSPECTION

E. THE FOLLOWING RESTRICTIONS ARE CONSIDERED NECESSARY FOR SAFE OPERATION: (Use attachment if necessary)

AIRCRAFT INSPECTION AND LOGBOOK ENTRY

F. CERTIFICATION – I hereby certify that I am the registered owner (or his agent) of the aircraft described above; that the aircraft is registered with the Federal Aviation Administration in accordance with Title 49 of the United States Code 44101 et seq. and applicable Federal Aviation Regulations; and that the aircraft has been inspected and is safe for the flight described.

DATE	NAME AND TITLE (Print or Type)	SIGNATURE
2-26-2008	JANE A. AVIATOR	*Jane A. Aviator*

VIII. AIRWORTHINESS DOCUMENTATION (FAA/DESIGNEE use only)

☐ A. Operating Limitations and Markings in Compliance with 14 CFR Section 91.9, as applicable.	☐ G. Statement of Conformity, FAA Form 8130-9 (Attach when required)
☐ B. Current Operating Limitations Attached	☐ H. Foreign Airworthiness Certification for Import Aircraft (Attach when required)
☐ C. Data, Drawings, Photographs, etc. (Attach when required)	☐ I. Previous Airworthiness Certificate Issued in Accordance with 14 CFR Section ______ CAR ______ (Original Attached)
☐ D. Current Weight and Balance information Available in Aircraft	
☐ E. Major Repair and Alteration, FAA Form 337 (Attach when required)	☐ J. Current Airworthiness Certificate Issued in Accordance with 14 CFR Section ______ (Copy Attached)
☐ F. This inspection Recorded in Aircraft Records	☐ K. Light-Sport Aircraft Statement of Compliance, FAA Form 8130-15 (Attach when required)

FAA Form 8130-6 (10-04) Previous Edition Dated 5/01 May be Used Until Depleted, except for Light-Sport Aircraft NSN: 0052-00-024-7006

Figure 5-1. *Form 8130-6* (page 2 of 2).

UNITED STATES OF AMERICA
DEPARTMENT OF TRANSPORTATION - FEDERAL AVIATION ADMINISTRATION
SPECIAL AIRWORTHINESS CERTIFICATE

A	CATEGORY/DESIGNATION	SPECIAL FLIGHT PERMIT	
	PURPOSE	MAINTENANCE	
B	MANUFAC TURER	NAME N/A	
		ADDRESS N/A	
C	FLIGHT	FROM SHAWNEE, OKLAHOMA	
		TO DOWNTOWN AIRPARK, OKLAHOMA CITY, OK	
D	N- 25565		SERIAL NO. 182-582672
	BUILDER CESSNA		MODEL C-182L
E	DATE OF ISSUANCE 03-01-96		EXPIRY 04-01-99
	OPERATING LIMITATIONS DATED 03-01-96 ARE A PART OF THIS CERTIFICATE		
	SIGNATURE OF FAA REPRESENTATIVE Darrel A. Freeman		DESIGNATION OR OFFICE NO. OKC-MIDO-41

Any alteration, reproduction or misuse of this certificate may be punishable by a fine not exceeding $1,000 or imprisonment not exceeding 3 years, or both. THIS CERTIFICATE MUST BE DISPLAYED IN THE AIRCRAFT IN ACCORDANCE WITH APPLICABLE FEDERAL AVIATION REGULATIONS.

FAA Form 8130-7 (10/82) *REVERSE SIDE OF APPLICATION OF AIRWORTHINESS CERTIFICATE*

Figure 5-2. *FAA Form 8130-7, Special Airworthiness Certificate.* The FAA issues FAA Form 8130-7, Special Airworthiness Certificate, as a special flight permit.

6 Light-Sport Aircraft

Light-sport aircraft (LSA) is a growing sector of the general aviation community, specific to the United States. The Federal Aviation Administration (FAA) promulgated sport pilot (SP)/LSA regulations in 2004. This significant change in Title 14 of the Code of Federal Regulations (14 CFR) allowed easier and lower-cost access to general aviation.

Definition

LSA as defined in 14 CFR part 1, section 1.1, "means an aircraft, other than a helicopter or powered-lift that, since its original certification, has continued to meet the following:

(1) A maximum takeoff weight of not more than—
  (i) 1,320 pounds (600 kilograms) for aircraft not intended for operation on water; or
  (ii) 1,430 pounds (650 kilograms) for an aircraft intended for operation on water.
(2) A maximum airspeed in level flight with maximum continuous power (V_H) of not more than 120 knots CAS under standard atmospheric conditions at sea level.
(3) A maximum never-exceed speed (V_{NE}) of not more than 120 knots CAS for a glider.
(4) A maximum stalling speed or minimum steady flight speed without the use of lift-enhancing devices (V_{S1}) of not more than 45 knots CAS at the aircraft's maximum certificated takeoff weight and most critical center of gravity.
(5) A maximum seating capacity of no more than two persons, including the pilot.
(6) A single, reciprocating engine, if powered.
(7) A fixed or ground-adjustable propeller if a powered aircraft other than a powered glider.
(8) A fixed or autofeathering propeller system if a powered glider.
(9) A fixed-pitch, semi-rigid, teetering, two-blade rotor system, if a gyroplane.
(10) A nonpressurized cabin, if equipped with a cabin.
(11) Fixed landing gear, except for an aircraft intended for operation on water or a glider.
(12) Fixed or retractable landing gear, or a hull, for an aircraft intended for operation on water.
(13) Fixed or retractable landing gear for a glider."

LSA Certification

Several different kinds of aircraft may be certificated as LSA. Airplanes (both powered and gliders), rotorcraft (gyroplanes only, not true for helicopters), powered parachutes, weight-shift control aircraft, and lighter-than-air craft (free balloons and airships) may all be certificated as LSA if they fall within weight and other guidelines established by the FAA.

LSA Registration

If you purchased a newly manufactured LSA that is to be certificated as an experimental LSA under 14 CFR part 21, section 21.191(i)(2), or a special LSA under 14 CFR part 21, section 21.190, then you must provide the following documentation to the FAA Civil Aviation Registry Aircraft Registration Branch (AFS-750):

- Aeronautical Center (AC) Form 8050-88 (as revised), Light-Sport Aircraft Manufacturer's Affidavit, or its equivalent, completed by the LSA manufacturer, unless previously submitted to AFS-750 by the manufacturer,
- Evidence of ownership from the aircraft manufacturer,
- AC Form 8050-1, Aircraft Registration Application, and
- Registration fee.

The FAA Light Sport Aviation Branch (AFS-610) or your local Flight Standards District Office (FSDO) can assist you with questions about LSA registration.

Available Resources

There are a number of resources available to assist LSA owners and operators.

Light Sport Aviation Branch, AFS-610

AFS-610 manages and provides oversight of the SP examiner and the LSA repairman-training programs, and also provides subject matter experts for FAA and the aviation industry concerning the SP/LSA aircraft safety initiatives. AFS-610 performs the following functions:

- Acceptance of LSA Repairman courses
- Light Sport Standardization Board
- Oversight of designated SP Examiners
- SP Examiner Initial Training Seminar
- SP Examiner Recurrent Training Program

AFS-610 contact information is available in the FAA Contact Information appendix on pages A1–A2 of this handbook.

Experimental Aircraft Association

The Experimental Aircraft Association (EAA) was founded in 1953 by a group of individuals in Milwaukee, Wisconsin, who were interested in building their own airplanes. Through the decades, the organization expanded its mission to include antiques, classics, warbirds, aerobatic aircraft, ultralights, helicopters, light sport, and contemporary manufactured aircraft.

EAA is an excellent resource for light sport aircraft owners and operators. You can contact the EAA at:

EAA Aviation Center
3000 Poberezny Rd
Oshkosh, WI 54902
(800) JOIN-EAA *phone*
www.eaa.org

Regulatory Guidance

Regulatory guidance for LSA owners and operators includes:

- FAA Order 8130.2 (as revised), Airworthiness Certification of Aircraft and Related Products
- FAA Order 8130.33 (as revised), Designated Airworthiness Representatives: Amateur-Built and Light-Sport Aircraft Certification Functions

7 Aircraft Maintenance

Maintenance means the preservation, inspection, overhaul, and repair of aircraft, including the replacement of parts. The purpose of maintenance is to ensure that the aircraft remains airworthy throughout its operational life. A properly maintained aircraft is a safe aircraft.

Although maintenance requirements vary for different types of aircraft, experience shows that most aircraft need some type of preventive maintenance every 25 hours or less of flying time, and minor maintenance at least every 100 hours. This is influenced by the kind of operation, climactic conditions, storage facilities, age, and construction of the aircraft. Maintenance manuals are available from aircraft manufacturers or commercial vendors with revisions for maintaining your aircraft.

While the requirements for maintaining your aircraft are contained in Title 14 of the Code of Federal Regulations (14 CFR), it is essential for every aircraft owner to remember that specific maintenance requirements are available from the aircraft manufacturer.

Maintenance Responsibilities

14 CFR part 91, section 91.403, places primary responsibility on the owner or operator for maintaining an aircraft in an airworthy condition. Certain inspections must be performed on your aircraft, and you must maintain the airworthiness of the aircraft between the required inspections by having any defects corrected. 14 CFR part 91, section 91.327 pertains to light-sport aircraft. Light-sport aircraft certificated in the light sport category under 14 CFR part 21, section 21.190 must be maintained by an FAA-certificated airframe and powerplant (A&P) mechanic or a light-sport repairman with a maintenance rating.

14 CFR Part 91, Subpart E

14 CFR part 91, subpart E, requires the inspection of all civil aircraft at specific intervals to determine the overall condition. The interval generally depends on the type of operations in which the aircraft is engaged. Some aircraft need to be inspected at least once every 12 calendar months, while inspection is required for others after each 100 hours of operation. In other instances, an aircraft may be inspected in accordance with an inspection system set up to provide for total inspection of the aircraft on the basis of calendar time, time in service, number of system operations, or any combination of these factors.

To determine the specific inspection requirements and rules for the performance of inspections, you should refer to 14 CFR part 91, subpart E, which prescribes the requirements for various types of operations.

Manufacturer Maintenance Manuals

All inspections must follow the manufacturer maintenance manual, including the instructions for continued airworthiness concerning inspection intervals, parts replacement, and life-limited items as applicable to your aircraft. The maintenance manuals provided by the manufacturer of your aircraft are your best available resource on issues of aircraft maintenance.

Preventive Maintenance

14 CFR lists 32 relatively uncomplicated repairs and procedures defined as preventive maintenance. Certificated pilots, excluding student and recreational pilots, may perform preventive maintenance on any aircraft owned or operated by them that are not used in air carrier service. These preventive maintenance operations are listed in 14 CFR Part 43, Appendix A, Preventive Maintenance. 14 CFR part 43 also contains other rules to be followed in the maintenance of aircraft.

Inspections

In order to provide a reasonable assurance that aircraft are functioning properly, the Federal Aviation Administration (FAA) requires a series of aircraft inspections somewhat similar to the many currency requirements for airmen. This section outlines the basic inspection requirements for aircraft.

Annual Inspection

Most general aviation aircraft require an annual inspection pursuant to 14 CFR part 91, section 91.409.

Excluded aircraft:

- Use an approved progressive inspection plan;
- Carry a special flight permit; or
- Carry a provisional airworthiness certificate.

The annual inspection must be completed and approved by a mechanic with an inspection authorization (IA) once every 12 calendar months. For example, if the aircraft's annual is endorsed on June 16, 2008, the next annual inspection is due before July 1, 2009; otherwise the aircraft may not be flown without authorization (e.g., a special flight or "ferry" permit).

A ferry permit is required to fly an aircraft that is out of annual, such as in the case of flying to another airport for the inspection. Chapter 5 discusses the issuance of special flight permits. You can contact your local Flight Standards District Office (FSDO) for instructions on applying for a special flight or ferry permit.

100-Hour Inspection

The 100-hour (14 CFR part 91, section 91.409) inspection is required for aircraft that either:

- Carry any person (other than a crewmember) for hire, or
- Are provided by any person giving flight instruction.

The 100-hour limit may be exceeded by 10 hours for the purposes of flying to a place where the inspection can be completed. The excess time must be included in computing the next 100 hours of time in service.

Some examples of "for hire" operations under 14 CFR part 91 that subject the aircraft to the 100-hour inspection requirement include:

- An aerial photography flight, or
- A flight instructor providing an aircraft, or any operation that supplies both flight instruction and an aircraft. (An aircraft provided by the (student) pilot receiving instruction is not subject to the 100-hour inspection.)

Condition Inspection

A condition inspection is required once every 12 calendar months for light-sport aircraft certificated in the light-sport category. In accordance with 14 CFR part 91, section 91.327, the condition inspection must be performed by "a certificated repairman (light-sport aircraft) with a maintenance rating, an appropriately rated mechanic, or an appropriately rated repair station in accordance with inspection procedures developed by the aircraft manufacturer or a person acceptable to the FAA."

Other Inspection Programs

The annual and 100-hour inspection requirements do not apply to large (over 12,500 pounds) airplanes, turbojets, or turbopropeller-powered multiengine airplanes, or to airplanes for which the owner or operator complies with the progressive inspection requirements. Details of these requirements may be determined by reference to 14 CFR part 43, section 43.11; 14 CFR part 91, subpart E; and by inquiry at the local FSDO.

Progressive Inspection

To minimize maintenance downtime, the owner may opt for a progressive inspection plan. Progressive inspections benefit owners whose aircraft experience high usage such as fixed base operators (FBOs), flight schools, and corporate flight departments. Unlike an annual inspection, a progressive inspection allows for more frequent but shorter inspection phases, only if all items required for the annual and 100-hour inspections are inspected within the required time. The authority to use a progressive inspection plan is non-transferable. Once the aircraft is sold, an annual becomes due within 12 calendar months of the last complete cycle. The 100-hour inspection is due at the completion of the next 100 hours of operation. Most airframe manufacturers provide a boilerplate progressive maintenance plan.

14 CFR Part 43, Appendix D, Scope and Detail of Items (as Applicable to the Particular Aircraft) To Be Included in Annual and 100-Hour Inspections, contains a list of general items to be checked during inspections.

Altimeter System Inspection

The aircraft's static system, altimeter, and automatic altitude-reporting (Mode C) system must have been inspected and tested in the preceding 24 calendar months before flying instrument flight rules (IFR) in controlled airspace. 14 CFR Part 43, Appendix E, Altimeter System Test and Inspection, lists the items that must be checked.

Transponder Inspection

The transponder must be inspected every 24 calendar months. 14 CFR Part 43, Appendix F, ATC Transponder Tests and Inspections, lists the items that must be checked. Additionally, the installation of or modification to a transponder must be inspected for data errors as well.

Preflight Inspection

A pilot is required to conduct a thorough preflight inspection before every flight to ensure that the aircraft is safe for flight. Pilots should review the maintenance status of the aircraft as a part of the preflight inspection.

Repairs and Alterations

All repairs and alterations of standard airworthiness certificated aircraft are classified as either major or minor. 14 CFR part 43, appendix A, describes the alterations and repairs considered major. Major repairs or alterations are approved for return to service on FAA Form 337, Major Repair and Alteration, by an appropriately rated certificated repair station, an FAA-certificated A&P mechanic holding an IA, or a representative of the Administrator. Minor repairs and minor alterations may be approved for return to service with a proper entry in the maintenance records by a certificated A&P mechanic or an appropriately certificated repair station.

Alterations to light-sport aircraft certificated in the light-sport aircraft category under 14 CFR part 21, section 21.190, must be authorized by the manufacturer or a person acceptable to the FAA in accordance with 14 CFR part 91, section 91.327.

Minimum Equipment List/Configuration Deviation List

If your aircraft has an approved Minimum Equipment List (MEL), the MEL should be used to determine if a flight may be initiated with inoperative aircraft equipment without the issuance of a special flight permit. Your Airplane Flight Manual (AFM) may also include a Configuration Deviation List (CDL) prepared by the manufacturer.

If your aircraft does not have an approved MEL, and you have inoperative equipment or instruments, then you must refer to 14 CFR part 91, section 91.213, to determine if a special flight permit is needed to operate the aircraft.

FAA Resources

You can find all of the maintenance requirements applicable to your aircraft in 14 CFR by accessing the relevant regulations on the FAA website at www.faa.gov. The best resource for answering questions about the maintenance necessary on your aircraft is your local FSDO.

Experimental Aircraft

If you make any major alterations to your experimental aircraft, you must notify your local FSDO of those alterations.

8 Maintenance Records

An aircraft owner is required to keep aircraft maintenance records for the airframe, engine, propeller, and appliances. These records must contain a description of the work performed on the aircraft, the date the work was completed, the certificated mechanic's signature, the type of Federal Aviation Administration (FAA) certificate, and the certificate number and signature of the person approving the aircraft for return to service.

Responsibilities of the Aircraft Owner

All recordkeeping is primarily the responsibility of the aircraft owner. The airframe and powerplant (A&P) mechanic is responsible for the work he or she performs. The owner of an aircraft must also ensure that maintenance personnel make appropriate entries in the aircraft maintenance records indicating the aircraft has been approved for return to service. The owner's aircraft records shall also contain the inspections required pursuant to Title 14 of the Code of Federal Regulations (14 CFR) part 91, section 91.409.

Proper management of aircraft operations begins with a good system of maintenance records. A properly completed maintenance record provides the information needed by the owner and maintenance personnel to determine when scheduled inspections and maintenance are to be performed.

Aircraft maintenance records must include:

- The total time in service of the airframe, each engine, and each propeller;
- The current status of life-limited parts of each airframe, engine, propeller, rotor, and appliance;
- The time since the last overhaul of all items installed on the aircraft, which are required to be overhauled on a specified time basis;
- The identification of the current inspection status of the aircraft, including the time since the last inspection required by the inspection program under which the aircraft and its appliances are maintained;
- The current status of applicable Airworthiness Directives (ADs) including, for each, the method of compliance, the AD number, and the revision date. If the AD involves recurring action, the time and date the next action is required; and
- A copy of the major alterations to each airframe, engine, propeller, and appliance.

These records are retained by the owner and are transferred with the aircraft when it is sold. *Figure 8-1* at the end of this chapter is a maintenance records checklist you can use to document compliance with the applicable maintenance requirements.

These records may be discarded when the work is repeated or superseded by other work, or 1 year after the work is performed.

⚠ CAUTION: Keep in mind that as a result of repairs or alterations, such as replacing radios and installing speed kits, it may be necessary to amend the weight and balance report, equipment list, flight manual, etc.

Logbooks

Most maintenance performed on an aircraft is recorded in the aircraft logbooks. As an aircraft owner, it is important to ensure that your aircraft's logbooks are complete and up to date at all times. The aircraft logbooks outline the maintenance history of your aircraft, and any prospective buyer will want to review the aircraft and all maintenance performed. In addition, any A&P or certificated repair station performing maintenance on your aircraft will want to review the prior maintenance performed on the aircraft.

Airworthiness Directives

An aircraft owner is required to comply with all applicable ADs issued by the FAA for his or her aircraft. The FAA issues ADs to notify aircraft owners and other interested persons of unsafe conditions and to specify the corrective action required, including conditions under which the aircraft may continue to be operated.

The aircraft's maintenance records should indicate the current status of all applicable ADs, including for each:

- AD number,
- Method of compliance,
- Revision date, and
- Recurring action (if applicable) including the time and date of the next action required.

Safety Directives

The owner or operator of an aircraft having a special airworthiness certificate must comply with each safety directive applicable to the aircraft that corrects an existing unsafe condition or corrects the condition in a manner different from safety directive specifications if the person issuing the directive agrees with the action. Otherwise, the owner or operator may, in accordance with 14 CFR part 91, section 91.327, "obtain an FAA waiver from the provisions of the safety directive based on a conclusion that the safety directive was issued without adhering to the applicable consensus standard."

Light-sport category aircraft certificated pursuant to 14 CFR part 21, section 21.190 have mandatory compliance with all manufacturer safety directives.

Service Bulletins

A service bulletin contains a recommendation from the manufacturer, with which that manufacturer believes the aircraft owner should comply, that often reflects a safety-of-flight issue that the manufacturer believes should be addressed within a certain timeframe. It may result from an improvement developed by the manufacturer, or it may address a defect in a product or published documentation.

The manufacturer responds to one of these situations by issuing a service bulletin that recommends a certain type of inspection, replacing certain components, performing maintenance in a specific manner, or limiting operations under specified conditions. Sometimes, compliance with a service bulletin may be triggered by the occurrence of a particular event (e.g., the lapse of time or operation under certain types of conditions).

FAA Form 337, Major Repair and Alteration

A mechanic who performs a major repair or major alteration shall record the work on FAA Form 337 and have the work inspected and approved by a mechanic who holds an inspection authorization (IA). (Light-sport aircraft do not require Form 337 when altering a non-approved product.) A signed copy shall be given to the owner and another copy sent to the FAA Aircraft Registration Branch (AFS-750) in Oklahoma City, Oklahoma within 48 hours of aircraft approval for return to service. However, when a major repair is done by a certificated repair station, the customer's work order may be used and a release given as outlined in 14 CFR Part 43, Appendix B, Recording of Major Repairs and Major Alterations. You can obtain additional information and instructions for completing FAA Form 337 in Advisory Circular (AC) 43-9 (as revised), Aircraft Maintenance Records. *Figure 8-2* at the end of this chapter is a sample FAA Form 337.

Entries into Aircraft Maintenance Records

Each time maintenance, including preventive maintenance, is performed on your aircraft, an appropriate entry must be added to the maintenance records.

14 CFR Part 43, Section 43.9, Content, form, and disposition of maintenance, preventive maintenance, rebuilding, and alteration records (except inspections performed in accordance with part 91, part 125, §135.411(a)(1), and §135.419 of this chapter

Any person who maintains, rebuilds or alters an aircraft, airframe, aircraft engine, propeller, or appliance shall make an entry containing:

- A description of the work or some reference to data acceptable to the FAA,
- The date the work was completed,
- The name of the person who performed the work, and
- If the work was approved for return to service, the signature, certificate number, and kind of certificate held by the person approving the aircraft for return to service.

14 CFR Part 43, Section 43.11, Content, form, and disposition of records for inspections conducted under parts 91 and 125 and §§135.411(a)(1) and 135.419 of this chapter

When a mechanic approves or disapproves an aircraft for return to service after an annual, 100-hour, or progressive inspection, an entry shall be made including:

- Aircraft time in service,
- The type of inspection,
- The date of inspection,
- The signature, certificate number, and kind of certificate held by the person approving or disapproving the aircraft for return to service, and
- A signed and dated listing of discrepancies and unairworthy items.

14 CFR Part 91, Section 91.409, Inspections

Inspection entries for 14 CFR part 91, section 91.409(e) airplanes over 12,500 pounds, turbo jet, or turbopropeller-powered multiengine airplanes are made according to 14 CFR part 43, section 43.9, and shall include:

- The kind of inspection performed,
- A statement by the mechanic that the inspection was performed in accordance with the instructions and procedures for the kind of inspection program selected by the owner, and
- A statement that a signed and dated list of any defects found during the inspection was given to the owner, if the aircraft is not approved for return to service.

14 CFR Part 91, Section 91.411, Altimeter system and altitude reporting equipment tests and inspections

14 CFR part 91, section 91.411, requires that every airplane or helicopter operated in controlled airspace under instrument flight rules (IFR) conditions have each static pressure system, each altimeter, and each automatic pressure altitude reporting system tested and inspected every 24 calendar months. The aircraft maintenance records must include:

- A description of the work,
- The maximum altitude to which the altimeter was tested, and
- The date and signature of the person approving the aircraft for return to service.

14 CFR Part 91, Section 91.413, ATC transponder tests and inspections

14 CFR part 91, section 91.413, requires that anyone operating an Air Traffic Control (ATC) transponder specified in 14 CFR part 91, section 91.215(a), have it tested and inspected every 24 calendar months. The aircraft maintenance records must include:

- A description of the work, and
- The date and signature of the person approving the airplane for return to service.

14 CFR Part 91, Section 91.207, Emergency locator transmitters

14 CFR part 91, section 91.207, requires that no person may operate a U.S. registered civil airplane unless there is attached to the airplane a personal type or an automatic type emergency locator transmitter (ELT) that is in operable condition and meets applicable requirements of Technical Standard Order (TSO)-C91.

⚠ CAUTION: New ELT installations after June 21, 1995, must meet TSO-C91A (the first revised, or amended, version).

Batteries used in ELT shall be replaced when:

- The transmitter has been in use for more than 1 cumulative hour, or
- 50 percent of the ELT's useful life has expired.

The expiration date for replacing the battery shall be legibly marked on the outside of the transmitter and entered in the aircraft maintenance records.

Amateur-Built Aircraft

The condition inspection for amateur-built aircraft replaces the annual inspection.

Available Resources

Your local FSDO can help you establish your aircraft maintenance program and the necessary maintenance records. Additional information relating to aircraft maintenance records can be obtained from:

- 14 CFR Part 39, Airworthiness Directives
- 14 CFR Part 43, Maintenance, Preventive Maintenance, Rebuilding, and Alteration
- 14 CFR Part 91, General Operating and Flight Rules
- AC 43-9 (as revised), Maintenance Records

These publications are available on the FAA website at www.faa.gov and from U.S. Government Printing Office (GPO) bookstores located throughout the United States. For more information about obtaining these publications, refer to the information contained in chapter 11.

Maintenance Records Checklist

STATUS	ITEM	NOTES
☐	100-Hour inspection	Keep records until the work is repeated or superseded by other work, or 1 year after the work is performed.
☐	Annual inspection	Keep records until the work is repeated or superseded by other work, or 1 year after the work is performed.
☐	Progressive inspections	Keep records until the work is repeated or superseded by other work, or 1 year after the work is performed.
☐	Other required or approved inspections (e.g., condition inspections/mandatory tracking of safety directives for light-sport aircraft)	Keep records until the work is repeated or superseded by other work, or 1 year after the work is performed.
☐	Total Time in Service (airframe, engine(s), propeller(s))	Records are retained by the owner and transferred with the aircraft when it is sold.
☐	Current status of life-limited parts (airframe, engine(s), propeller(s), rotor, and appliances)	Records are retained by the owner and transferred with the aircraft when it is sold.
☐	Time since last overhaul of all items installed on the aircraft (required to be overhauled on a specified time basis)	Records are retained by the owner and transferred with the aircraft when it is sold.
☐	Identification of current inspection status of the aircraft (including time since last inspection required by the inspection program under which aircraft and appliances are maintained)	Records are retained by the owner and transferred with the aircraft when it is sold.
☐	Current status of applicable ADs (including method of compliance, the AD number, and the revision date)	Records are retained by the owner and transferred with the aircraft when it is sold.
☐	Copy of current major alterations (airframe, engine, propeller, and appliances)	Records are retained by the owner and transferred with the aircraft when it is sold.

Figure 8-1. *Maintenance Records Checklist.* This checklist includes the types of information that should be kept with your aircraft's maintenance records.

US Department of Transportation
Federal Aviation Administration

MAJOR REPAIR AND ALTERATION
(Airframe, Powerplant, Propeller, or Appliance)

Form Approved OMB No. 2120-0020 11/30/2007	Electronic Tracking Number
For FAA Use Only	

INSTRUCTIONS: Print or type all entries. See Title 14 CFR §43.9, Part 43 Appendix B, and AC 43.9-1 (or subsequent revision thereof) for instructions and disposition of this form. This report is required by law (49 U.S.C. §44701). Failure to report can result in a civil penalty for each such violation. (49 U.S.C. §46301(a))

1. Aircraft	Nationality and Registration Mark: N114AZ	Serial No.: 18259223	
	Make: CESSNA	Model: 182L	Series:
2. Owner	Name *(As shown on registration certificate)*: O & W INC.	Address *(As shown on registration certificate)*: Address 1888 CIRRUS AVENUE; City OKLAHOMA CITY; State OK; Zip 73405; Country U.S.	

3. For FAA Use Only

SAMPLE

4. Type		**5. Unit Identification**			
Repair	Alteration	Unit	Make	Model	Serial No.
[X]	[ ]	AIRFRAME	————	*(As described in Item 1 above)*	————
[ ]	[ ]	POWERPLANT			
[ ]	[ ]	PROPELLER			
[ ]	[ ]	APPLIANCE	Type Manufacturer		

6. Conformity Statement

A. Agency's Name and Address	B. Kind of Agency		C. Certificate No.
Name: KATHY P. AILERON	X	U. S. Certificated Mechanic	Manufacturer
Address: 411 GULFSTREAM DRIVE		Foreign Certificated Mechanic	C. Certificate No.
City: OKLAHOMA CITY State: OK		Certificated Repair Station	A&P 122234566
Zip: 73125 Country: U.S.		Certificated Maintenance Organization	

D. I certify that the repair and/or alteration made to the unit(s) identified in item 5 above and described on the reverse or attachments hereto have been made in accordance with the requirements of Part 43 of the U.S. Federal Aviation Regulations and that the information furnished herein is true and correct to the best of my knowledge.

Extended range fuel per 14 CFR Part 43 App. B [ ]	Signature/Date of Authorized Individual: Kathy P. Aileron MARCH 23,2002

7. Approval for Return to Service

Pursuant to the authority given persons specified below, the unit identified in item 5 was inspected in the manner prescribed by the Administrator of the Federal Aviation Administration and is [X] Approved [ ] Rejected

BY						
		FAA Flt. Standards Inspector		Manufacturer		Maintenance Organization	Persons Approved by Canadian Department of Transport
		FAA Designee		Repair Station	X	Inspection Authorization	Other *(Specify)*

Certificate or Designation No.	Signature/Date of Authorized Individual
233346566	Ed Mechanic MARCH 26, 2002

FAA Form 337 (10-06)

Figure 8-2. *FAA Form 337, Major Repair and Alteration.* You can obtain instructions for completing FAA Form 337 on the FAA website at www.faa.gov or from your local FSDO.

8

NOTICE

Weight and balance or operating limitation changes shall be entered in the appropriate aircraft record. An alteration must be compatible with all previous alterations to assure continued conformity with the applicable airworthiness requirements.

8. Description of Work Accomplished
(If more space is required, attach additional sheets. Identify with aircraft nationality and registration mark and date work completed.)

N114AZ	MARCH 26, 2002
Nationality and Registration Mark	Date

1. Removed right wing from aircraft and removed skin from outer 6 feet. Repaired buckled spar 49 inches from tip in accordance with attached photographs and figure 1 of drawing dated March 23, 2002.

   Date: March 26, 2002, inspected splice in Item 1 and found it to be in accordance with data indicated. Splice is okay to cover. Inspected internal wing assembly for hidden damage and condition.

   Ed Mechanic

   Ed Mechanic, A&P 233346566 IA

2. Primed interior wing structure and replaced skin P/N's 63-0085, 63-0086, and 63-00878 with same skin 2024-T3, .025 inches thick. Rivet size and spacing all the same as original and using procedures in Chapter 2, Section 3, of AC 43.13-1B CHG 1, dated 2001.
3. Replaced stringers as required and installed 6 splices as per attached drawing and photographs.
4. Installed wing, rigged aileron, and operationally checked in accordance with manufacturer's maintenance manual.
5. No change in weight or balance.

END

☐ Additional Sheets Are Attached

FAA Form 337 (10-06)

Figure 8-2. *FAA Form 337, Major Repair and Alteration* (Page 2 of 2).

EMERGENCY AIRWORTHINESS DIRECTIVE

U.S. Department of Transportation
Federal Aviation Administration

Aircraft Certification Service
Washington, DC

We post Emergency ADs on the internet at www.faa.gov/aircraft/safety/alerts.

March 12, 2008

is sent to all owners and operators

series reciprocating

9 Airworthiness Directives

An Airworthiness Directive (AD) is an important tool used by the Federal Aviation Administration (FAA) to communicate unsafe operating conditions relating to aircraft and aircraft equipment to aircraft owners. A primary safety function of the FAA is to require the correction of unsafe conditions found in an aircraft, aircraft engine, propeller, rotor, or appliance when such conditions exist or are likely to exist or develop in other products of the same design. These unsafe conditions can exist because of a design defect, maintenance, or other causes.

Title 14 of the Code of Federal Regulations (14 CFR) Part 39, Airworthiness Directives, defines the authority and responsibility of the Administrator in requiring the necessary corrective action to address unsafe conditions. ADs are used to notify aircraft owners and other interested persons of unsafe conditions and to specify the conditions under which the product may continue to be operated.

Types of ADs Issued

The FAA issues two categories of ADs:

- Normal Issue
- Emergency Issue

Standard AD Process

The standard AD process is to issue a Notice of Proposed Rulemaking (NPRM), followed by a Final Rule. After an unsafe condition is discovered, a proposed solution is published in the Federal Register as an NPRM, which solicits public comment on the proposed action. After the comment period closes, the final rule is prepared, taking into account all substantive comments received, with the rule perhaps being changed as warranted by the comments. The preamble to the Final Rule AD provides response to the substantive comments or states there were no comments received.

Emergency AD

In certain cases, the critical nature of an unsafe condition may warrant the immediate adoption of a rule without prior notice and solicitation of comments. The intent of an Emergency AD is to rapidly correct an urgent safety of flight situation. This is an exception to the standard process. If time by which the terminating action must be accomplished is too short to allow for public comment (that is, less than 60 days), then a finding of impracticability is justified for the terminating action, and it can be issued as an immediately adopted rule. The immediately adopted rule will be published in the Federal Register with a request for comments. The Final Rule AD may be changed later if substantive comments are received.

Superseded AD

An AD is no longer in effect when it is superseded by a new AD. The superseding AD identifies the AD that is no longer in effect. There are no compliance requirements for a superseded AD.

Compliance with ADs

For purposes of compliance, ADs may be divided into two categories:

- Those of an emergency nature requiring immediate compliance before further flight, or
- Those of a less urgent nature requiring compliance within a relatively longer period of time.

ADs are the "final rule" and compliance is required unless specific exemption is granted. Aircraft owners are responsible for ensuring compliance with all pertinent ADs. This includes those ADs that require recurrent or continuing action. For example, an AD may require a repetitive inspection each 50 hours of operation, meaning the particular inspection must be accomplished and recorded every 50 hours of time in service.

⚠ CAUTION: Aircraft owners are reminded that there is no provision to overfly the maximum hour requirement of an AD unless it is specifically written into the AD.

Amateur-Built Aircraft

For help in determining if an AD applies to your amateur-built aircraft, contact your local Flight Standards District Office (FSDO).

Summary of ADs

14 CFR part 91, section 91.417, requires a record to be maintained that shows the current status of applicable ADs, including:

- Method of compliance;
- AD number and revision date;
- Date and time when due again, if recurring;
- Certified mechanic's signature;
- Type of certificate; and
- Certificate number of the repair station or mechanical performing the work.

For ready reference, many aircraft owners keep a chronological listing of the pertinent ADs in the back of their aircraft and engine records. Generally, a summary of ADs contains all the valid ADs previously published. *Figure 9-1* is a sample form of summary of ADs.

Obtaining ADs

Both AD categories (small and large aircraft) are published in biweekly supplements. All ADs are available on the FAA website at www.faa.gov. Advisory Circular (AC) 39-7 (as revised), Airworthiness Directives, provides additional guidance and information for aircraft owners and operators about their responsibilities for complying and recording ADs. For more information, contact the FAA Regulatory Support Division, Delegation and Airworthiness Programs Branch (AIR-140). AIR-140 contact information is available in the FAA Contact Information appendix on pages A1–A2 of this handbook.

AIRWORTHINESS DIRECTIVE COMPLIANCE RECORD

Aircraft: PA-22-135 N2631A S/N 22-903

Engine: Lycoming 0-290-D2 S/N 4563-21

Propeller: Sensenich M 76AM2 S/N 6662

AD & AMEND NUM.	REV. NUM. & DATE	SUBJECT	DATE/ HOURS AT COMP.	METHOD OF COMP.	ONE TIME	RECUR-RING	COMP. DUE DATE/HRS	AUTHORIZED SIGNATURE & NUMBER
76-07-12 39-3024	R-1 8-30-77	Bendix ignition switch	11-11-94 1850TT	Operational check and inspection	X			Phil Lomax A&P 000000000
93-18-03 39-8688	Original 10-29-93	One-piece venturi	3-17-95 1850OTT	Installed one-piece venturi Carb S/N BR-549		X	1900TT	Phil Lomax A&P 000000000

Figure 9-1. *Sample Airworthiness Directives Compliance Record.* This sample AD Compliance Record is intended to show you an acceptable format for recording the required information to evidence your aircraft's compliance with applicable ADs. The FAA does not prescribe a specific format; however, the information discussed in this chapter must be maintained with the aircraft's maintenance records.

10 Service Difficulty Program

The Service Difficulty Program is an information system designed to provide assistance to aircraft owners, operators, maintenance organizations, manufacturers, and the Federal Aviation Administration (FAA) in identifying aircraft problems encountered during service.

Background

The Service Difficulty Program provides for the collection, organization, analysis, and dissemination of aircraft service information to improve service reliability of aeronautical products. The primary sources of this information are aircraft maintenance facilities, owners, and operators. The incentive for early detection is to expedite corrective actions and ultimate solutions, thereby minimizing the effect of equipment failure on safety.

Each problem reported contributes to the improvement of aviation safety through the identification of a potential problem area and the alerting of other persons to it. This focusing of attention on a problem has led to improvements in the design and maintainability of aircraft and aircraft products.

Advisory Circular (AC) 20-109 (as revised), Service Difficulty Program (General Aviation), describes the Service Difficulty Program as it applies to general aviation. It also includes instructions for completing FAA Form 8010-4, Malfunction or Defect Report. The information from these reports is compiled and published as Maintenance Alerts.

By pooling everyone's knowledge about a situation, the FAA can detect mechanical problems early enough to correct them before they might possibly result in accidents/incidents which should make flying safer, more enjoyable, and certainly less expensive.

FAA Form 8010-4, Malfunction or Defect Report

General aviation aircraft service difficulty information is normally submitted to the FAA by use of FAA Form 8010-4. However, information will be accepted in any form or format when FAA Form 8010-4 is not readily available for use.

The information contained in the FAA Form 8010-4 is stored in a computerized data bank for retrieval and analysis. Items potentially hazardous to flight are telephoned directly to the Regulatory Support Division, Aviation Data Systems Branch (AFS-620) personnel by FAA aviation safety inspectors (ASIs) in local Flight Standards District Offices (FSDOs). These items are immediately referred to, and promptly handled by, the appropriate FAA offices.

Aircraft owners, pilots, and mechanics are urged to report all service problems promptly, using FAA Form 8010-4 or any other form or format. You may obtain a copy of the form from any FSDO. No postage is required.

FAA Form 8010-4 is also available in a fillable PDF format on the FAA website at http://forms.faa.gov/forms/faa8010-4.pdf for mailing, or it can be submitted electronically on the FAA website at www.faa.gov. *Figure 10-1* at the end of this chapter is a sample FAA Form 8010-4.

Maintenance Alerts

The FAA publishes AC 43-16 (as revised), Aviation Maintenance Alerts, monthly on its website to provide the aviation community with a means for interchanging service difficulty information and sharing information on aviation service experiences.

Background

The Maintenance Alert program leads to improved aeronautical product durability, reliability, and safety. The articles contained in the Maintenance Alerts are derived from the Malfunction or Defect Reports submitted by aircraft owners, pilots, mechanics, repair stations, and air taxi operators.

Maintenance specialists review the reports and select pertinent items for publication in the Maintenance Alerts. The information is brief and advisory and compliance is not mandatory. However, the information is intended to alert you to service experience, and, when applicable, direct your attention to the manufacturer's recommended corrective action.

Accessing Maintenance Alerts

You can access current and back issues of this publication on the FAA website at http://www.faa.gov/aircraft/safety/alerts/aviation_maintenance/, which allows free access to each month's Maintenance Alerts.

Contact Information

You can contact the FAA Aviation Systems Data Branch (AFS-620) regarding the Maintenance Alert Program. AFS-620 contact information is available in the FAA Contact Information appendix on pages A1–A2 of this handbook.

DEPARTMENT OF TRANSPORTATION
FEDERAL AVIATION ADMINISTRATION

MALFUNCTION OR DEFECT REPORT

		OPER. Control No.	
		ATA Code	N404DH
		1. A/C Reg. No.	N-
Enter pertinent data	MANUFACTURER	MODEL/SERIES	SERIAL NUMBER
2. AIRCRAFT	Beechcraft	C-35	D-3311
3. POWERPLANT	Continental	E-225	30904
4. PROPELLER	Hartzell	HCA2U20 4A1	AK-710

5. SPECIFIC PART *(of component)* CAUSING TROUBLE

Part Name	MFG. Model or Part No.	Serial No.	Part/Defect Location.
Tube-fuel pressure	35-924126	N/A	B-nut Carb end

6. APPLIANCE/COMPONENT *(Assembly that Includes part)*

Comp/Appl Name	Manufacturer	Model or Part No.	Serial Number
N/A	N/A	N/A	N/A

Part TT	Part TSO	Part Condition	7. Date Sub.
4100	N/A	Cracked	4-28-05

8. Comments *(Describe the malfunction or defect and the circumstances under which it occurred. State probable cause and recommendations to prevent recurrence.)*

During a local flight, a fuel odor was apparent. Flight was terminated at the local airport with a normal landing. After engine shutdown and exiting the aircraft the pilot observed fuel leaking from the lower engine cowl. Investigation revealed an aluminum fuel pressure gauge line cracked at the B nut where it attached to a carburetor fitting (see attached drawing). Line appears to be original (40+ yrs.). A combination of age and vibration may have caused the crack. Recommend checking line every 100 hrs. and replacing as necessary.

Optional Information:

Check a box below, if this report is related to an aircraft

☐ Accident; Date ________ ☐ Incident; Date ________

DISTRICT OFFICE | OTHER | COMMUTER | FAA | MFG. | AIR TAXI | MECH. | OPER. | REP STA.

OPERATOR DESIGNATION

SUBMITTED BY: David Waterski

TELEPHONE NUMBER (405) 555 — 4316

FAA FORM 8010-4 (10-92) SUPERSEDES PREVIOUS EDITIONS

Use this space for continuation of Block 8 *(if required)*.

SAMPLE

FUEL PRESSURE GAUGE LINE

CRACKED LINE AT "B" NUT

PAPERWORK REDUCTION ACT STATEMENT: The information collected on this form is used to evaluate certification standards, maintenance programs, and regulatory requirements. The information is required to ensure safety in air transportation. It is estimated that it will take approximately 9 minutes to complete the form. Providing this information is mandatory. No assurance of confidentiality is given. Please note that an agency may not conduct or sponsor, and a person is not required to respond to,a collection of information unless it displays a currently valid OMB control number. The control number for this collection of information is 2120-0003. Comments concerning the accuracy of this burden and suggestions for reducing the burden should be directed to the FAA at: 800 Independence Ave SW, Washington, DC 20591, Attn: Information Collection Clearance Officer, ABA-20

Figure 10-1. *FAA Form 8010-4, Malfunction or Defect Report.* You can obtain instructions for completing FAA Form 8010-4 on the FAA website at www.faa.gov or from your local FSDO.

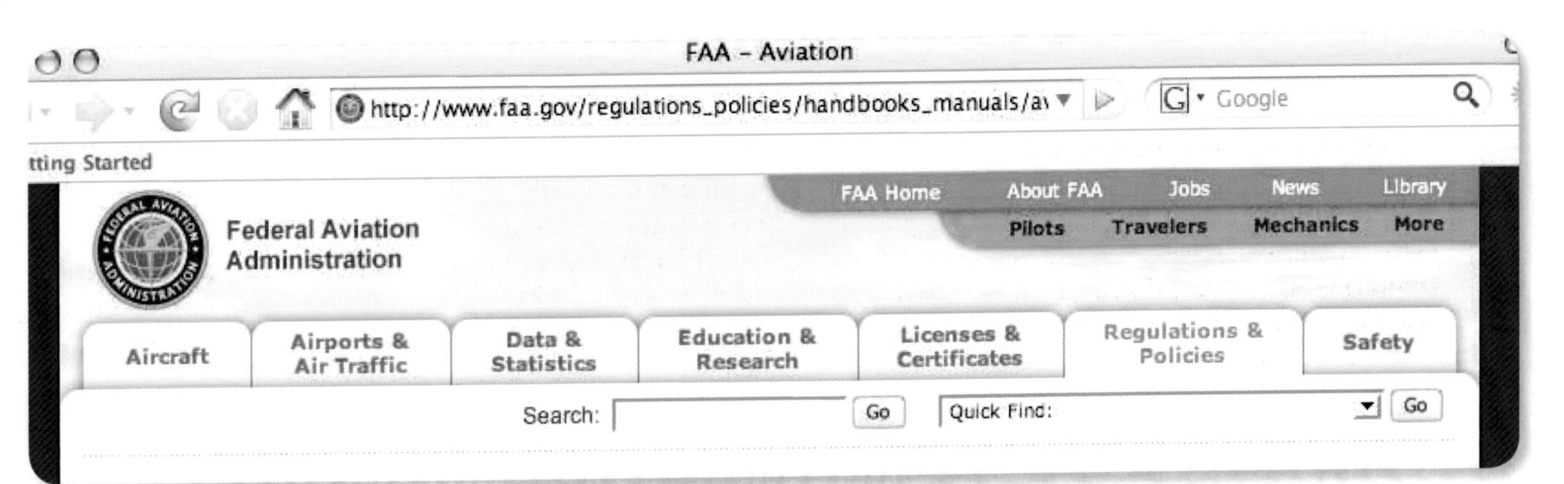

11 Obtaining FAA Publications and Records

There are several ways to obtain various Federal Aviation Administration (FAA) publications. The easiest way to locate a particular FAA document and/or Title 14 of the Code of Federal Regulations (14 CFR) part is on the FAA website at www.faa.gov. In addition, you can order some FAA publications directly from the FAA or from the Government Printing Office (GPO). The FAA Records Checklist in *Figure 11-1* at the end of this chapter is a list of addresses to assist you in obtaining the publications and records discussed in this chapter.

FAA publications and supporting regulatory guidance material fall into several specific categories, which are explained in this chapter. If you cannot find the information you are seeking, refer to the FAA website, which contains a great deal of useful information for aircraft owners and operators.

Advisory Circulars

The FAA issues an Advisory Circular (AC) to inform the aviation public, in a systematic way, of nonregulatory material of interest. The content of an AC is not binding on the public unless it is incorporated into a regulation by reference.

AC 00-2 (as revised) , Advisory Circular Checklist, contains a list of current FAA ACs and provides detailed instructions on obtaining copies. It also contains a list of U.S. GPO bookstores throughout the United States that stock many Government publications. This AC may be accessed on the FAA website at www.faa.gov by selecting "Regulations & Policies" from the main menu bar.

Airworthiness Directives

The FAA uses an Airworthiness Directive (AD) to notify aircraft owners and other interested persons of unsafe conditions and to specify the conditions under which the product may continue to be operated. You can access ADs on the FAA website at www.faa.gov by selecting "Regulations & Policies" from the main menu bar.

You can subscribe to ADs at the FAA Regulatory & Guidance Library (RGL) website, http://rgl.faa.gov. Current and historical ADs are also available on the RGL website.

Temporary Flight Restrictions

A Temporary Flight Restriction (TFR) is a geographically limited, short term, airspace restriction, typically in the United States. TFRs often encompass major sporting events, natural disaster areas, air shows, space launches, and Presidential movements. You can access current TFRs on the FAA website at www.faa.gov by selecting "Regulations & Policies" from the main menu bar.

⚠ CAUTION: You should check the current TFRs every time you fly.

Notice to Airmen

You can obtain the most recent Notice to Airmen (NOTAMs) from the FAA website at www.faa.gov by selecting "Regulations & Policies" from the main menu bar. Current NOTAMs are also available from Flight Service Stations at 1-800-WX-BRIEF (1-800-992-7433).

14 CFR

While aircraft owners and operators are responsible for compliance with all applicable 14 CFR parts, you may find the following parts most relevant to your aircraft operations:

- 14 CFR Part 1, Definitions and Abbreviations
- 14 CFR Part 21, Certification Procedures for Products and Parts
- 14 CFR Part 23, Airworthiness Standards: Normal, Utility, Acrobatic, and Commuter Category Airplanes
- 14 CFR Part 33, Airworthiness Standards: Aircraft Engines
- 14 CFR Part 35, Airworthiness Standards: Propellers
- 14 CFR Part 39, Airworthiness Directives
- 14 CFR Part 43, Maintenance, Preventive Maintenance, Rebuilding, and Alteration
- 14 CFR Part 45, Identification and Registration Marking
- 14 CFR Part 47, Aircraft Registration
- 14 CFR Part 49, Recording of Aircraft Titles and Security Documents
- 14 CFR Part 61, Certification: Pilots, Flight Instructors, and Ground Instructors
- 14 CFR Part 65, Certification: Airmen Other than Flight Crewmembers
- 14 CFR Part 91, General Operating and Flight Rules

AC 00-44 (as revised), Status of Federal Aviation Regulations, contains the current status of the 14 CFR parts, including changes issued, price list, and ordering instructions. This AC may be obtained on the FAA website at www.faa.gov.

Handbooks and Manuals

The FAA publishes a series of handbooks and manuals designed for aircraft, general aviation enthusiasts, examiners, and inspectors. You can find most of these handbooks and manuals, including *Plane Sense*, on the FAA website at www.faa.gov by selecting "Regulations & Policies" from the main menu bar.

Aircraft Records

The FAA Civil Aviation Registry Aircraft Registration Branch (AFS-750) maintains registration records on individual aircraft and serves as a warehouse for airworthiness documents received from FAA field offices. You can access information on requesting aircraft records, as well as current fee information, on the FAA website at www.faa.gov by selecting "Licenses & Certificates" from the main menu bar.

Request Aircraft Records

You can mail or fax your request for aircraft records to AFS-750, or submit your request online.

Written requests from outside the United States must include a check or money order (in U.S. funds) payable to the FAA to cover the projected fee. You can obtain the correct fee by contacting AFS-750. AFS-750 contact information is available at the end of this chapter in *Figure 11-1*, FAA Records Checklist, and in the FAA Contact Information appendix on pages A1–A2 of this handbook.

Format

You can order paper copies or electronic copies (on CD-ROM) of aircraft records. Each CD-ROM contains one aircraft record. You can view the CD-ROM files using Adobe Acrobat Reader. The FAA includes a copy of the latest reader on the CD-ROM.

Most records for aircraft removed from the U.S. Civil Aircraft Register before 1984 are in storage and available only in paper format. You can ask the FAA to retrieve these records.

Airman Records

The FAA Civil Aviation Registry Airmen Certification Branch (AFS-760) maintains airman records. You can access information on requesting airman records, as well as current fee information, on the FAA website at www.faa.gov by selecting "Licenses & Certificates" from the main menu bar.

There are two ways to obtain copies of your airman records. You can mail the FAA Aeronautical Center (AC) Form 8060-68, Request for Copies of My Complete Airman File; or a signed, written request stating your name, date of birth, and social security number or certificate number. *Figure 11-2* at the end of this chapter is a sample FAA Form 8060-68.

You can also have copies of your airman records released to a third party. Requests for airman records can be mailed to AFS-760. AFS-760 contact information is available at the end of this chapter in *Figure 11-1*, FAA Records Checklist, and in the FAA Contact Information appendix on pages A1–A2 of this handbook.

When the FAA receives your request, they will notify you of the total charges due and payment options.

For an airman or third party to obtain copies of medical records or a duplicate medical certificate, the Freedom of Information Act (FOIA) request should be mailed to the FAA Civil Aerospace Medical Certification Division, Medical Certification Branch (AAM-331). AAM-331 contact information is available at the end of this chapter in *Figure 11-1*, FAA Records Checklist, and in the FAA Contact Information appendix on pages A1–A2 of this handbook.

FAA RECORDS CHECKLIST

STATUS	RECORDS	CONTACT INFORMATION
☐	Aircraft Documents	Federal Aviation Administration Aircraft Registration Branch, AFS-750 P.O. Box 25504 Oklahoma City, OK 73125 (405) 954-3116
☐	Airman Records (Replacement Certificate)	Federal Aviation Administration Airmen Certification Branch, AFS-760 P.O. Box 25082 Oklahoma City, OK 73125 (405) 954-3261
☐	Airman Medical Records	Federal Aviation Administration Aeromedical Certification Branch, AAM-331 P.O. Box 26080 Oklahoma City, OK 73126-5063 (405) 954-4821
☐	FOIA Desk (Third Party Request, Duplicate Medical Certificate)	Federal Aviation Administration Aeromedical Certification Branch, AAM-331 Attention: FOIA Desk P.O. Box 26200 Oklahoma City, OK 73125-9914

Figure 11-1. *FAA Records Checklist.* The easiest way to locate information on obtaining records from the FAA is on the FAA website at www.faa.gov. You can also use this checklist to determine which FAA branch to contact regarding the records requested.

U.S. DEPARTMENT OF TRANSPORTATION
Federal Aviation Administration
AIRMEN CERTIFICATION BRANCH, AFS-760

REQUEST FOR COPIES OF MY COMPLETE AIRMAN FILE

PRIVACY ACT: This information is required under the authority of Transportation Title 49 U.S.C. Section 44703 et. seq. Your request cannot be processed unless the data below is complete. Disclosure of your Social Security Number (SSN) and/or date of birth (DOB) is optional. Refusal to furnish your SSN and/or DOB will not result in the denial of any right, benefit, or privilege provided by law; however, failure to provide the SSN and/or DOB may result in the delay of a response or the processing of your inquiry. Routine uses of records maintained in the system include; categories of users and the purpose of such uses i.e., to determine that airmen are certified in accordance with the provision of the Federal Aviation Regulations; repository of documents used by individuals and potential employers to determine validity of airmen qualifications; to support investigative efforts of Federal, State, and local law enforcement agencies; supportive information in court cases concerning individual status and/or qualifications in law suits; to provide data for the Comprehensive Airmen Information System.

WILLIAM THOMAS WRIGHT
Full Name (As it appears on the certificate/Please print)

9-12-53 (Date-of-Birth) WICHITA, KANSAS (Place-of-Birth)

111-22-3333
(Social Security No., Certificate No., Class of Certificate)

341 PIPER ROAD
(Street Address, Apt./Suite No., PO Box/Rural Route No.)

OKLAHOMA CITY (City) OK (State) 73125 (Zip Code)

FEES: The fees for these copies are $2 for Search of Records, $3 for Certification of a file, 25 cents for the first page, and 5 cents for each additional page. Upon receipt of the requested complete airman file, you will be notified of the total charges due and the options of payment. **Please allow 6 to 8 weeks for processing.**

William Thomas Wright
Signature (Typed or Printed signature is not acceptable)

4-28-08
Date

SAMPLE

Mail this request to:
Federal Aviation Administration
Airmen Certification Branch, AFS-760
PO Box 25082
Oklahoma City, OK 73125-0082

Medical Records, Accident and Incident Reports, and Violation Information are not part of an airman file. To request these, please contact the appropriate office below:

For Medical or combined Student/Medical, Please contact:

Federal Aviation Administration
Medical Certification Branch, AAM-331
Post Office Box 26200
Oklahoma City, OK 73125-9914

For Accidents, Incidents, or Violation Information, Please contact:

Federal Aviation Administration
Aviation Data System Branch, AFS-620
Post Office Box 25082
Oklahoma City, OK 73125-0082

AC 8060-68 (12/06)

Figure 11-2. *FAA Form 8060-68, Request for Copies of My Complete Airman File.* You can obtain instructions for completing FAA Form 8060-68 on the FAA website at www.faa.gov or from your local FSDO.

Appendix A: FAA Contact Information

The information contained in this appendix will help you contact the appropriate Federal Aviation Administration (FAA) office.

FAA Office	Page(s)
U.S. Department of Transportation **Federal Aviation Administration** 800 Independence Avenue, SW Washington, DC 20591-0004 (866) TELL-FAA (866-835-5322) *toll-free* www.faa.gov	vii
Federal Aviation Administration **Airman Testing Standards Branch, AFS-630** P.O. Box 25082 Oklahoma City, OK 73125-0082 (405) 954-4151 *phone* afs630comments@faa.gov	iii
Federal Aviation Administration **Aircraft Registration Branch, AFS-750** P.O. Box 25504 Oklahoma City, OK 73125-0504 (866) 762-9434 *toll-free* (405) 954-3116 *phone* (405) 954-3548 or (405) 954-8068 *fax*	2-2, 2-3, 4-1, 4-3, 4-4, 4-5, 11-2, 11-4
Federal Aviation Administration **Light Sport Aviation Branch, AFS-610** P.O. Box 25082 Oklahoma City, OK 73125-0082 (405) 954-6400 *phone* (405) 954-4104 *fax*	2-4, 4-5, 6-2
Federal Aviation Administration **Delegation & Airworthiness Programs Branch, AIR-140** P.O. Box 26460 Oklahoma City, OK 73125-4902 (405) 954-4103 *phone*	9-2

A

FAA Office	Page(s)
Federal Aviation Administration **Aviation Systems Data Branch, AFS-620** ATTN: AFS-620 ALERTS P.O. Box 25082 Oklahoma City, OK 73125-0082 (405) 954-4391 *phone*	10-2
Federal Aviation Administration **Airmen Certification Branch, AFS-760** P.O. Box 25082 Oklahoma City, OK 73125-0082 (405) 954-3261 *phone*	11-3, 11-4
Federal Aviation Administration **Aeromedical Certification Branch, AAM-331** P.O. Box 26080 Oklahoma City, OK 73126-5063 (405) 954-4821 *phone*	11-3, 11-4
Federal Aviation Administration **Aeromedical Certification Branch, AAM-331** ATTN: FOIA Desk P.O. Box 26200 Oklahoma City, OK 73125-9914 (405) 954-4821 *phone*	11-4

Appendix B: Regulatory Guidance Index

The information contained in this appendix will help you locate regulatory guidance information including pertinent Code of Federal Regulations parts, FAA Orders, and Advisory Circulars.

FAA Orders (as revised)	Page(s)

Advisory Circulars (ACs) (as revised)	Page(s)

Notes

Notes

Notes

Notes

Notes

Notes

Notes

Notes

Notes

Notes

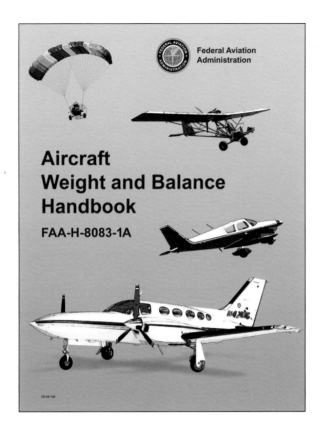

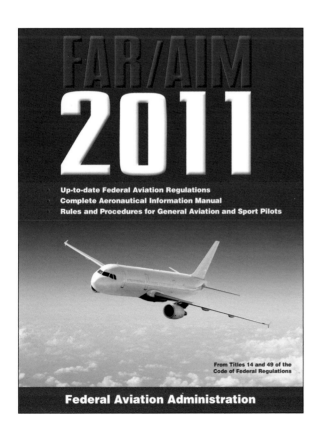

ALSO AVAILABLE

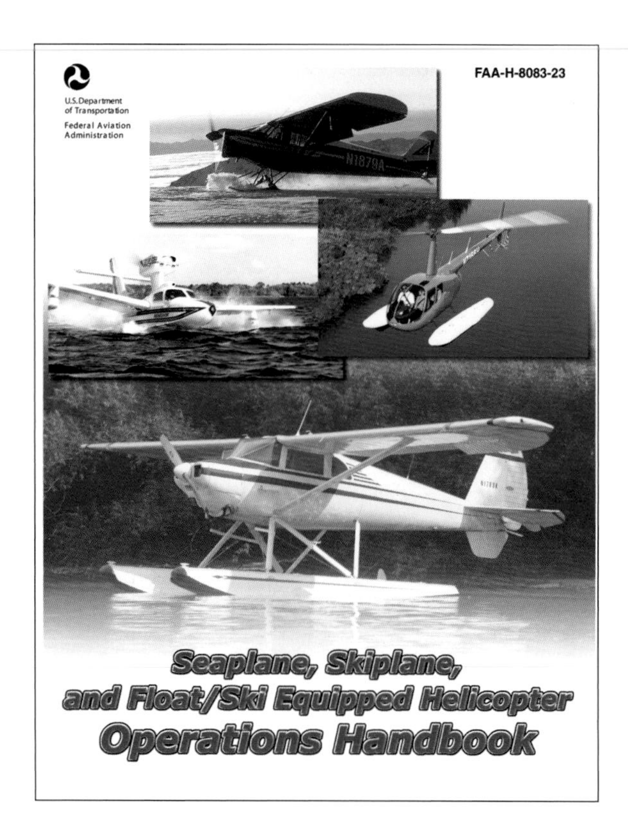

Seaplane, Skiplane and Float/Ski Equipped Helicopter Operations Handbook: FAA-H-8083-23-1

The ultimate guide to water-related aircraft piloting.

Federal Aviation Administration

This comprehensive handbook provides the most up-to-date, definitive information on piloting water-related aircraft. Along with full-color photographs and illustrations, detailed descriptions make complicated tasks easy to understand while the index and glossary provide the perfect references for finding any topic and solving any issue.

The FAA leaves no question unanswered in the most complete book on how to fly water-related aircraft available on the market. The *Seaplane, Skiplane, and Float/Ski Equipped Helicopter Operations Handbook* is the perfect addition to the bookshelf of all aircraft enthusiasts, FAA fans, and novice and experienced pilots alike.

$12.95 Paperback • 96 pages • March 2011

Powered Parachute Flying Handbook: FAA-H-8083-29

From the FAA, the only handbook you need to learn to fly a powered parachute.

Federal Aviation Administration

As far back as the twelfth century, people have loved to parachute. From China's umbrella and Leonardo da Vinci's pyramid-shaped flying device to the first airplane jump in 1912, the urge to leap and soar with the wind has long been a part of history. Parachuting has come a long way since its earliest days due to the advancement of technology, and now more people than ever are taking up this incredible sport. With the *Powered Parachute Flying Handbook* you can make your flying ambitions a reality.

$24.95 Paperback • 160 pages

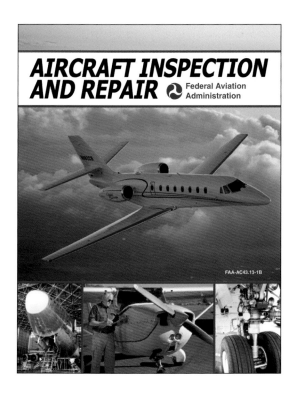

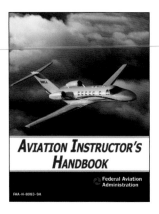

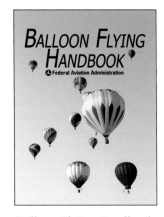

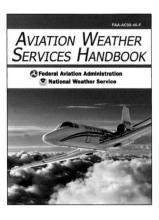

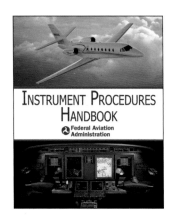